PROBING THE
NEW SOLAR SYSTEM

PROBING
THE NEW
SOLAR SYSTEM

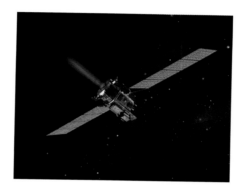

JOHN WILKINSON

CSIRO
PUBLISHING

National Library of Australia Cataloguing-in-Publication entry
Wilkinson, John.

Probing the new solar system / John Wilkinson.
9780643095755 (pbk.)
Includes index.
Bibliography.

Solar system
Outer space – Exploration
Astronomy

523.2

Published by
CSIRO PUBLISHING
150 Oxford Street (PO Box 1139)
Collingwood VIC 3066
Australia

Telephone: +61 3 9662 7666
Local call: 1300 788 000 (Australia only)
Fax: +61 3 9662 7555
Email: publishing.sales@csiro.au
Web site: www.publish.csiro.au

Front cover images by NASA

Set in ITC New Baskerville
Edited by Elaine Cochrane
Text design by James Kelly
Cover design by James Kelly, based on a concept by NASA
Typeset by J&M Typesetting
Printed in China by 1010 Printing International Ltd

CSIRO PUBLISHING publishes and distributes scientific, technical and health science books, magazines and journals from Australia to a worldwide audience and conducts these activities autonomously from the research activities of the Commonwealth Scientific and Industrial Research Organisation (CSIRO).

The views expressed in this publication are those of the author(s) and do not necessarily represent those of, and should not be attributed to, the publisher or CSIRO.

CONTENTS

INTRODUCTION

The overwhelming importance of the solar system lies in the fact that we are part of it; its origin and evolution are part of our own history. Astronomers have observed the solar system for the past few centuries via telescopes from the Earth's surface. Then, in 1957, a new method of exploration began with the launch of the first artificial satellite – humans had entered the Space Age. Since this time, humans have improved the technology of their spacecraft to the point where they can now send probes deep into the solar system to places never seen before. In the past few decades there have been space probes sent to explore the crater-strewn surface of Mercury, and the roasting hot surface of Venus. Humans have even walked on the surface of the Moon. We have placed robotic probes on the surface of Mars and used them to search for life on this red planet. The giant planets Jupiter and Saturn, together with their many moons and ring systems, have also undergone extensive exploration by space probes. Saturn's rings are arguably the flattest structure in the solar system, but from end to end they would reach from Earth to the Moon. The cold icy planets of Uranus and Neptune also have ring systems, and more moons than previously thought. In the past few years we have also discovered other planet-like bodies beyond Neptune and Pluto, in far-out regions of the solar system called the Kuiper belt and Oort cloud. These new discoveries have made astronomers re-examine their understanding of the solar system.

This exploration has revealed that the Earth's planetary neighbours are fascinating worlds. Today we stand on the threshold of the next phase of planetary exploration. Many new missions are currently underway, and many more are being planned.

This book explores recent changes to our understanding of the solar system, in particular the effect on this understanding of the International Astronomical Union's 2006 revised definition of a planet. This book is therefore also a record of the many discoveries made about the solar system in recent years using space technology.

ACKNOWLEDGEMENTS

The author and publisher are grateful to the following for the use of photographs in this publication.

National Aeronautics and Space Administration (NASA)

European Space Agency

CSIRO Archives

While every care has been taken to trace and acknowledge copyright, the author apologies in advance for any accidental infringement where copyright has proved untraceable. He will be pleased to come to a suitable arrangement with the rightful owner in each case.

CHAPTER 1

WHY A NEW SOLAR SYSTEM?

THIS CHAPTER REVIEWS our early and current understandings of the solar system, the discovery of the planets, and some of the earlier space probes used in the exploration of space.

In the last few years our understanding of the solar system has changed. For thousands of years, humans have been fascinated by the movement of the stars and planets across the night sky. They have wondered what these objects are made of, how they move across the sky, and whether these worlds contain other living beings like us.

In ancient times people noted the position of the Sun in the various seasons and its effect on crop growth. They also knew how the Moon affected the tides. And they observed objects called planets moving against a background of stars. The Babylonians even developed a calendar based on the movement of the planets visible to the unaided eye. In fact, the names of the days of our week originate from the Sun, Moon, Mercury, Venus, Mars, Jupiter and Saturn. These objects are the classical objects of our night sky.

The word 'planet' comes from the Greek word meaning 'wanderers'. The Greeks observed that the planets wandered against a background of stars that remained relatively fixed in relation to each other. The band

Figure 1.1 Stars develop everywhere we look in space. In our region of the universe they form mostly in the arms of spiral galaxies. The solar system we live in is part of the Milky Way galaxy. The Milky Way has a spiral structure like that of the galaxy shown in this photograph taken by the Hubble Space Telescope. (Photo: NASA)

across the sky through which the planets moved was called the **zodiac**. The star groups or constellations that form the zodiac were given names, such as those of animals; for example, with a bit of imagination, the constellation Leo resembled a lion and Taurus resembled a bull.

Early Western and Arab civilisations and the ancient Greeks believed that the Earth was at the centre of the universe, with the Sun, Moon and the then known planets orbiting around it. This view was challenged by Polish astronomer Nicolaus Copernicus in the sixteenth century, when he suggested that all the planets, including the Earth, orbited the Sun in near-circular orbits. By using a Sun-centred model, Copernicus was able to determine which planets were closer to the Sun than the Earth, and which were further away. Because Mercury and Venus were always close to the Sun, Copernicus concluded that their orbits must lie inside that of the Earth. The other planets known at that time, Mars, Jupiter and Saturn, were often seen high in the night sky, far away from the Sun, so Copernicus concluded that their orbits must lie outside the Earth's orbit.

It was not until early in the seventeenth century that the German astronomer Johannes Kepler showed that the orbits of the planets around the Sun were elliptical, rather than circular. Kepler also showed that a planet moved faster when closer to the Sun and more slowly when further from the Sun, and he developed a mathematical relationship between the planet's distance from the Sun and the length of time it takes to orbit the Sun once. These three proven observations became known as Kepler's laws of planetary motion.

With the invention of the telescope in 1608, an Italian, Galileo Galilei, was able to gather data to support Copernicus's model for the Sun and planets. Galileo discovered four moons orbiting the planet Jupiter; he also observed sunspots moving across the surface of the Sun and craters on the Moon. Galileo's discovery that the planet Venus had phases just like Earth's Moon confirmed that Venus orbited the Sun closer than Earth and provided support for Copernicus's Sun-centred model.

One major problem restricting the full acceptance of Kepler's and Galileo's theories was that it was not known what kept the planets in orbit. People did not know how planets, once they started orbiting the Sun, could keep moving. Isaac Newton proposed the explanation of this motion in the seventeenth century. Newton put forward the idea that the Sun must be exerting a force on the planets to keep them in orbit. This force was called gravity, and it exists between any two masses (such as a planet and a star like our Sun). Using his law of gravity, Newton was able to prove the validity of Kepler's three laws of planetary motion. Newton also developed a universal law of gravitation, which states:

> *Two bodies attract each other with a force that is directly proportional to the product of their masses and inversely proportional to the square of the distance between them.*

This law means that the more mass a planet or star has, the greater its gravitational pull. This pull decreases with increasing distance from the object.

DISCOVERING NEW PLANETS

Towards the end of the eighteenth century, only six planets were known – Mercury, Venus, Earth, Mars, Jupiter and Saturn. In 1781, the British astronomer William Herschel accidentally discovered the seventh planet, Uranus. In 1846, Urbain Leverrier in France, and John Adams in England,

Figure 1.2   Astronomers use optical and radio telescopes to explore the universe. Pictured is the Parkes Radio Telescope in Australia. (Photo: CSIRO Archives)

independently used Newton's gravitational laws to predict that variations in the orbit of Uranus were caused by the influence of an eighth planet. Soon after, astronomers at the Berlin Observatory found the predicted planet and named it Neptune. In the early twentieth century, Percival Lowell and William Pickering predicted that another planet should exist beyond Neptune. In 1930, Clyde Tombaugh found a body, which was named Pluto, close to where Lowell and Pickering predicted it to be. Between 1930 and 2006 Pluto was regarded as the ninth planet of the solar system. However, in 2006 a meeting of the International Astronomical Union (IAU) decided on a definition of a planet that excluded Pluto, instead classing it as a 'dwarf planet', along with some other newly discovered bodies. Then, in 2008, the new category of 'plutoid' was established for Pluto and similar bodies. As a result we now have what many call 'the new solar system'.

What is a planet?

Traditionally, a planet has been regarded as a spherical body that orbits a star and is visible because it reflects light. The spherical shape is only possible when the object has enough mass for gravity to be able to pull it into a spherical shape. All planets, and many large moons and large asteroids, are spherical.

In August 2006 the IAU decided on the following definition of a planet:

To be a **planet** a body must:

1. be in orbit around the Sun,
2. have sufficient mass for self-gravity to overcome rigid body forces so that it assumes a hydrostatic equilibrium (nearly round) shape, and
3. have cleared the neighbourhood around its orbit.

What made a definition suddenly critical was the discovery of a number of objects beyond Pluto in the outer solar system. Before the new definition was accepted, there could have been as many as 50 objects classed as planets orbiting the Sun.

The IAU introduced two new classifications: 'dwarf planet' and 'plutoid'. A **plutoid** is a body that is in orbit around the Sun beyond Neptune, has sufficient mass for self-gravity to have pulled it into a spherical shape, has not cleared the neighbourhood around its orbit, and is not a satellite. A **dwarf planet** is one that orbits more closely around the Sun, has sufficient mass for self-gravity to have pulled it into a spherical shape, has not cleared the neighbourhood around its orbit, and is not a satellite. All other objects orbiting the Sun are collectively known as 'small solar system bodies'. Currently, there are a number of bodies regarded as plutoids, and more are expected to be added to the list over the next few years when more is known about them. Examples of plutoids include Pluto and Eris, which is further out from the Sun than Pluto. The former asteroid Ceres is the only dwarf planet. Plutoids and dwarf planets are not considered to be true planets mainly because they have not cleared their orbital path of other material.

As of mid-2008, there are eight major planets in the solar system. In order of distance from the Sun they are Mercury, Venus, Earth, Mars,

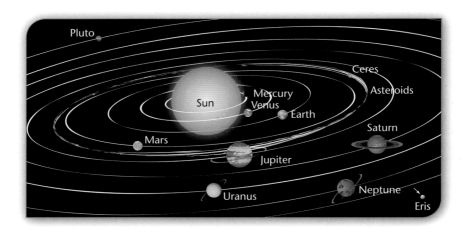

Figure 1.3   Major bodies in the solar system (not to scale).

Jupiter, Saturn, Uranus and Neptune. There are also one dwarf planet (Ceres) and three plutoids (Pluto, Makemake and Eris), but more bodies may be classed as plutoids in the future.

Many of the planets in the solar system have **natural satellites** or **moons** orbiting them. The planets Jupiter and Saturn have the most moons. True moons are large enough for gravity to have pulled their mass into a spherical shape, while smaller moons do not have enough mass and gravity and are irregular in shape. To be a moon, a body must be naturally orbiting a planet and be smaller than the planet. Most moons have been captured by planets during the formation of the solar system.

The four largest planets (Jupiter, Saturn, Uranus, Neptune) are also orbited by planetary rings of varying size and complexity. These rings are composed primarily of dust or particulate matter. The origin of such rings is not known, but they may be leftover debris from moons that have been torn apart by tidal forces.

Table 1.1 Major planets, dwarf planets and plutoids in the solar system (as of 2008)

Name	Class	Average distance from Sun (AU)[a]	Diameter (km)	Number of moons	Ring system
Mercury	major planet	0.38	4880	0	none
Venus	major planet	0.72	12 104	0	none
Earth	major planet	1.00	12 756	1	none
Mars	major planet	1.52	6794	2	none
Ceres	dwarf planet	2.76	950	0	none
Jupiter	major planet	5.20	142 984	63	faint
Saturn	major planet	9.54	120 536	62	prominent
Uranus	major planet	19.2	51 118	27	faint
Neptune	major planet	30.1	49 532	13	faint
Pluto	plutoid	39.5	2320	3	none
Eris	plutoid	60–100	2400	1	none
Makemake	plutoid	38–53	1500	0	none

a   Distances are given in astronomical units (AU) where one AU is the distance between the Earth and the Sun.

THE SOLAR SYSTEM – A FAMILY

The Sun, the planets and their moons form a family of bodies called the **solar system**. We now know that the Sun is indeed at the centre of the solar system, and that the major planets orbit the Sun in nearly circular orbits. Our understanding of the solar system has changed dramatically over the centuries as bigger and better telescopes were developed and more data on planetary motions were collected. Although people knew the planets of

Figure 1.4   The Orion Nebula is the nearest large site of star formation. (Photo: NASA)

the solar system existed, little was known about the nature of these worlds until space probes containing scientific instruments were sent to explore these objects.

The Sun fits the definition of a star because it emits its own heat and light through the process of thermonuclear fusion. Planets differ from stars in that they do not emit their own heat and light because they do not have enough mass for fusion to occur in their core. We see planets because they reflect sunlight. Stars and planets are spherical in shape, and are held in that shape by gravity.

The planets orbit the Sun in much the same plane, and because of this they all appear to move across the sky through a narrow band called the zodiac. Observed from a position above the Sun's north pole, all the planets orbit the Sun in an anti-clockwise direction. These orbits are often described as elliptical, but for most of the planets these ellipses are close to being circular (Figure 1.5). **Eccentricity** is a measure of how far an orbit deviates from circularity.

The time taken by a planet to orbit the Sun is called its **period** of revolution and is also the length of its year. A planet's year depends on its distance from the Sun: the further a planet is from the Sun, the slower its speed and the longer its year.

Planets also rotate or spin on an **axis**, which is an invisible line through their centre, from their north pole to their south pole. A planet's rotation period is known as its **day**. Planets also have varying degrees of **axial tilt**.

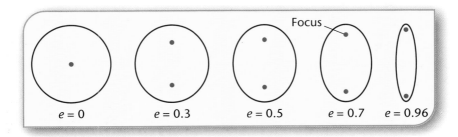

Figure 1.5   Shapes of planetary orbits. (e = eccentricity.)

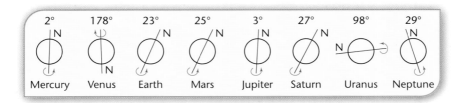

Figure 1.6   Axial tilt of each planet (approximate).

Axial tilt is the angle between a planet's axis of rotation and the vertical (Figure 1.6).

Each planet has its own gravitational field, which pulls objects towards it. Gravitational field strength is measured in newtons per kilogram at the surface of a planet. It takes a lot of energy to overcome gravity and escape from the surface of a planet. The minimum speed that an object (such as a rocket) must attain in order to travel into space from the surface of a planet, moon or other body is called the **escape velocity**. If the rocket's velocity is too low, gravity will pull it back down.

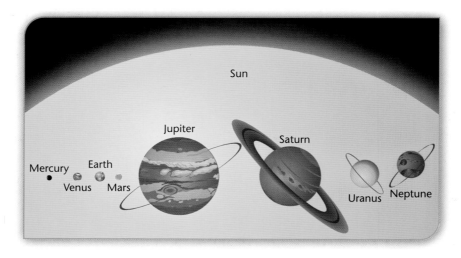

Figure 1.7   The planets in front of a disc of the Sun, with all bodies drawn to the same scale.

Planets also have a density, which is a measure of the amount of mass in a given volume. Density is measured in kilograms per cubic metre but is often quoted in grams per cubic centimetre as well.

The four 'inner planets' (Mercury, Venus, Earth and Mars) are called terrestrial planets. They are smaller, denser and rockier than the outer 'gas planets' (Jupiter, Saturn, Uranus and Neptune). The inner planets are also warmer, and rotate more slowly than the outer planets. The outer planets are gaseous planets, containing mostly hydrogen and helium with some methane and ammonia. These rapidly rotating planets are cold and icy with deep atmospheres.

During the 1800s, astronomers discovered a large number of small, rocky bodies orbiting the Sun between Mars and Jupiter. Bodies such as Ceres, Pallas and Vesta, which had been thought of as small planets for almost half a century, were reclassified with the new designation 'asteroid'.

Formation of the solar system

The solar system is thought to have formed about 4.5 billion years ago from a vast cloud of very hot gas and dust called the **solar nebula**. This cloud of interstellar material began to condense under its own gravitational forces. As a result, density and pressure at the centre of the nebula began to increase, producing a dense core of matter called the **protosun**. Collisions between the particles in the core caused the temperature to rise deep inside the protosun.

The planets and other bodies in the solar system formed because the solar nebula was rotating. Without rotation, everything in the nebula would have collapsed into the protosun. The rotating material formed a flat disc with a warm centre and cool edges. This explains why nearly all the planets now rotate in the same direction and in much the same plane. Astronomers have found similar discs around other stars.

As the temperature inside the protosun increased, light gases like hydrogen and helium were forced outward, while heavy elements remained closer to the core. The heavier material condensed to form the inner planets (which are mainly rock containing silicates and metals), while the lighter, gaseous material (methane, ammonia and water) condensed to form the outer planets. Thus a planet's composition depends on what material was available at different locations in the rotating disc and the temperature at each location.

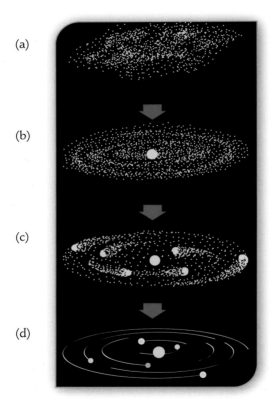

Figure 1.8   Stages in the formation of the solar system:
(a) A slowly rotating cloud of interstellar gas and dust begins to condense under its own gravity.
(b) A central core begins to form a protosun. A flattened disc of gas and dust surrounds the protosun, and begins to rotate and flatten.
(c) The planets begin to condense out of the flattened disc as it rotates.
(d) The planets have cleared their orbit of debris.

The formation of the solar system took millions of years. During this time the temperature and pressure of the protosun continued to increase. Finally the centre of the protosun became hot enough for nuclear fusion reactions to begin, and the Sun was born.

Stars like our Sun can take 100 million years to form from a nebula. Data from the oldest radioactive material in our solar system suggests it is about 4.6 billion years old.

Much of the debris leftover from the formation of the solar system is in orbit around the Sun in two regions, the **asteroid belt** and the **Kuiper belt**. The asteroid belt lies between Mars and Jupiter, while the Kuiper belt is a region beyond Neptune.

The asteroid belt contains over a million rocky bodies. About 2000 of these have very elliptical orbits, some of which cross Earth's orbit. Stony fragments ejected during asteroid collisions are called **meteoroids**. Some of these meteoroids enter the Earth's atmosphere and burn up, releasing light; such bodies are called **meteors**. Large meteors that impact with the ground form craters like those seen on the surface of the Moon.

The Kuiper belt is a bit like the asteroid belt, except that it is much further from the Sun and contains thousands of very cold bodies made of ice and rock. Objects in this outer region take up to 200 years to orbit the Sun.

There are other smaller objects on the outer edge of the solar system, in a region called the **Oort cloud**. This roughly spherical cloud also contains many objects left over from the formation of the solar system. **Comets** are icy bodies that originate from these outer regions. Many comets have highly elongated orbits that occasionally bring them close to the Sun. When this happens the Sun's radiation vaporises some of comet's icy material, and a long tail is seen extending from the comet's head. Each time they pass the Sun, comets lose about one per cent of their mass. Thus comets do not last forever. Comets eventually break apart, and their fragments give rise to many of the meteor showers we see from Earth.

Figure 1.9   Comet Hale-Bopp passed close to Earth in 1997. (Photo: H. Mikuz and B. Kambic, Crni Vrh Observatory, Slovenia)

PIONEERS OF SPACE EXPLORATION

Exploration of space became a reality on 4 October 1957, when Russia launched **Sputnik 1**, the first artificial satellite or human-made space probe to orbit Earth. By today's standards, Sputnik 1 was primitive. This satellite was little more than a spherical metal ball with four antennae, a radio transmitter and batteries. It was not equipped to take photographs. It orbited Earth once about every 95 minutes at a speed of 29 000 km/h. Sputnik 1 fell back to Earth on 4 January 1958. On 3 November 1957, **Sputnik 2** was launched. This was a more sophisticated satellite and carried a dog named Laika into orbit. The dog died in space. Russia launched nine much larger Sputniks between November 1957 and March 1961.

The Americans entered the space race on 31 January 1958, when they launched a satellite called **Explorer 1** into orbit via a US army rocket as part of the Mercury program. On 29 July 1958, America established the National Aeronautics and Space Administration (NASA).

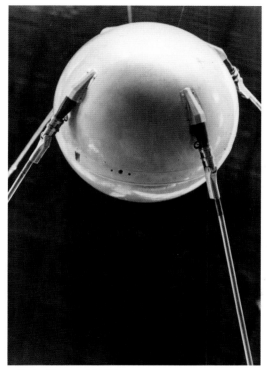

Figure 1.10   Sputnik 1. (Photo: NASA)

The first person to orbit Earth in a spacecraft was Russian Yuri Gagarin, on 12 April 1961. His flight aboard **Vostok 1** made one orbit of the Earth. The first American into space was Alan Shepard, on 5 May 1961. Shepard's single-seat capsule, called **Freedom 7**, made only a 15-minute flight before splashing down into the ocean. It was not until 20 February 1962 that John Glenn became the first US astronaut to complete a full orbit of the Earth. Glenn orbited three times in five hours in his space capsule called **Friendship 7**. Friendship 7 was also part of the Mercury program, which included six manned flights from 1961 to mid-1963. Mercury spacecraft were tiny and lightweight and could carry only one astronaut at a time; they were intended to test spacecraft technology and the effects of spaceflight on humans.

DID YOU KNOW?

John Glenn, the first US astronaut to orbit Earth, could have been America's first space fatality. During re-entry, the heat shield on Glenn's Friendship 7 capsule became loose. If the shield had fallen away completely, Glenn would have burned like a meteorite as the capsule plunged through the Earth's atmosphere. NASA ordered Glenn to delay releasing a retro-rocket pack attached to the shield. Engineers thought that the pack would help deflect the heat generated during re-entry. They were right, and his capsule splashed down into the ocean safely.

Glenn retired in 1964 and became a politician. He was elected to the US Senate in 1974. In 1998, Glenn, aged 77, became the oldest person in space when he flew on the space shuttle Discovery as part of a study on ageing.

The first woman into space was Soviet cosmonaut Valentina Tereshkova, on 16 June 1963. Her flight on **Vostok 6** lasted almost three full days. The first multi-member spacecraft to orbit Earth was **Voskhod 1**, launched on 12 October 1964 – this craft contained three Soviet cosmonauts. On 23 March 1965 the NASA spacecraft **Gemini 3** carried two US astronauts into orbit.

The first person to walk in space was Soviet cosmonaut Alexei Leonov (18 March 1965). Leonov floated around the outside of his spacecraft (**Voskhod 2**) for 20 minutes, linked by a gold-plated tether. Three months later, on 3 June 1965, US astronaut Edward White stepped out of his **Gemini 4** capsule.

Gemini program

Between 1964 and 1966, the USA undertook 10 manned flights as part of the Gemini program. The Gemini spacecraft could carry two astronauts and were intended to test spacecraft and equipment that would lead to putting a person on the Moon.

On 3 June 1965, US astronaut Edward White became the first American to walk in space. He took his spacewalk during the flight of **Gemini 4**. White spent 23 minutes floating in space, using a hand-held compressor to manoeuvre himself around.

On 16 March 1966, the crew of **Gemini 8** performed the world's first successful docking in space, linking their craft with an unmanned orbiting target vehicle. However, this mission also produced America's first real space emergency. Shortly after docking, a thruster jammed and the Gemini spacecraft began to spin violently. Undocking the target vehicle caused the Gemini capsule to spin even faster – 360 degrees every second. The two astronauts on board, Neil Armstrong and Dave Scott, struggled to gain control, and they had to make an emergency landing in the Pacific Ocean.

Figure 1.11   The meeting of Gemini 6 and Gemini 7, 185 miles (300 kilometres) above the Earth, on 16 December 1965. (Photo: NASA)

Soyuz spacecraft

In 1967 the Russians began their Soyuz missions into space. Soyuz spacecraft were designed to carry up to three cosmonauts, however **Soyuz 1** only carried one person, Vladimir Komarov, who died when his craft crashed during landing.

Early Moon probes

Many early space probes were directed at the Moon since it was the closest body to Earth and was a relatively easy target. The first spacecraft launched by the USA and aimed at the Moon (Pioneers 0, 1, and 3) failed to reach escape velocity or exploded. In 1959, **Luna 1** was launched by the USSR and became the first craft to fly past the Moon – it discovered the solar wind, and is now in solar orbit. America's **Pioneer 4** made a distant fly-by of the Moon in 1959. The first spacecraft to strike the Moon's surface was the USSR's **Luna 2** in September 1959. In the same year, **Luna 3** sent back the first photographs of the far side of the Moon. The first US probe to strike the Moon was **Ranger 4**, in 1962, but its camera failed to return any pictures. **Ranger 5** (October 1962) was to be a lander, but it became a fly-by because of a spacecraft failure. **Luna 9** made a soft landing on the Moon on 3 February 1966, and returned the first photographs from the lunar surface, while **Zond 5** (USSR) in 1968 was the first probe to orbit the Moon and return to a soft landing on Earth.

The USA's **Surveyor 5** landed on the Moon and sent back information on the composition of the lunar soil in 1967, while in the same year **Surveyor 6** landed on the Moon and successfully took off from the surface. In September 1970, the USSR's **Luna 16** became the first probe to return lunar soil samples to Earth.

NASA's Apollo program began in 1968 and was designed with the aim of eventually putting a person on the Moon. There were eleven manned Apollo flights, each of which was part of one of the greatest technological achievements by humans.

The Apollo spacecraft were lifted into space via Saturn V rockets. Each craft had the general shape of a cone, and they were launched with the narrow end pointing up to reduce air resistance during the flight through the atmosphere. They descended back through the atmosphere with the broad end pointing in the direction of travel to reduce speed and deflect the heat generated on re-entry.

Tragedy struck the Apollo program in January 1967, when a fire in their capsule killed the crew of Apollo 1 during a launch rehearsal. As a result Apollos 2 and 3 were cancelled, and missions 4, 5 and 6 were reassigned to automated test launches.

The first manned Apollo flight (**Apollo 7**) occurred on 11 October 1968. During 11 days orbiting Earth, the crew of Apollo 7 tested equipment and manoeuvres. On 21 December that same year, **Apollo 8**, with astronauts Frank Borman, James Lovell and William Anders on board, orbited the Moon 10 times before returning safely to Earth.

Figure 1.12   Apollo astronaut walking on the Moon. (Photo: NASA)

In May 1969, the crew of **Apollo 10** orbited the Moon and returned to Earth. They tested the lunar landing module, separating it from the command and service module and descending to within 50 000 feet (15 000 m) of the lunar surface. The astronauts took a large number of 70 mm photographs of the lunar surface.

The first lunar landing took place on 20 July 1969, when Neil Armstrong, followed by Edwin ('Buzz') Aldrin, walked on the Moon's surface in an area known as the Mare Tranquillitatis (Sea of Tranquillity). Their **Apollo 11** mission included Michael Collins, who was aboard the command module that remained in orbit above the Moon's surface. It had taken seven years of intensive planning, four manned test missions, and the combined effort of 400 000 engineers to get a human to the Moon. The astronauts returned to Earth with 21 kilograms of lunar rock material for examination.

Apollo 12 was a manned lunar landing that took place on 19 November 1969, in an area known as Oceanus Procellarum (Ocean of Storms). This area was also the landing site for the unmanned Surveyor 3 in 1967. Astronauts Conrad and Bean from Apollo 12 were able to retrieve portions of Surveyor 3, including a camera.

Apollo 13 lifted off for the Moon on 11 April 1970, but an explosion on board cut short the mission. The explosion destroyed both power and

DID YOU KNOW?

Edwin Aldrin, the second astronaut to walk on the Moon, weighed about 70 kg on Earth. His backpack weighed almost as much again, yet on the Moon he weighed only 22 kg. This is because the Moon's gravity is one-sixth that of Earth's. Aldrin felt like he was floating as he walked on the Moon's surface. Each step he took launched him into the air for seconds. To compensate for the mass in his backpack, Aldrin had to lean forward and he found it difficult to tell when he was standing upright.

propulsion systems of the command service module. The astronauts had to move into the lunar module in order to survive the return trip to Earth.

Astronauts Shepard and Mitchell from **Apollo 14** landed on the Moon on 5 February 1971. They collected 42 kg of rock samples from Fra Mauro crater, and used a hand-held cart to transport rocks and equipment. Apollo 14 commander Alan Shepard famously hit two golf balls on the Moon's surface.

On 26 July 1971, **Apollo 15** became the first mission on which astronauts used a lunar rover (a car-like vehicle) to move around on the Moon's surface. Batteries powered the rover and it ventured 12 km from the main landing craft during a seven-hour drive. It is still on the Moon today. Astronauts on **Apollo 16** also used a lunar rover to drive over the lunar surface during a 71-hour stay in 1972.

Figure 1.13   The lunar rover on the Moon. (Photo: NASA)

The Apollo Moon program ended in 1972 with **Apollo 17**. This spacecraft landed near the Descartes crater in the previously unvisited lunar highlands region. The last Apollo astronaut to walk on the Moon was Eugene Cernan, who saluted the US flag placed in the lunar soil. The Apollo missions ended with 12 astronauts having walked on the Moon, and 382 kilograms of lunar rock being returned to Earth. NASA originally planned 10 Apollo missions, but only six were completed. The last gasps of the Apollo program were the Skylab space station and the Apollo–Soyuz mission (a rendezvous between American and Soviet astronauts in Earth orbit).

THE FIRST SPACE STATIONS

A **space station** is a craft that is placed in stable orbit around Earth. Such crafts can be used to monitor space and Earth and provide stop-off points for future manned missions to planets.

Russia launched the world's first space station, **Salyut 1**, in 1971. Salyut was about 15 metres long with three compartments that housed dining and recreational areas, food and water storage, a toilet, control stations, and exercise and scientific equipment. The first crew lived aboard Salyut for 23 days, but were killed during their return to Earth. In subsequent years a total of seven Salyut space stations were placed in orbit. On 2 October 1984, three Soviet cosmonauts returned from **Salyut 7** after spending a record 237 days in space.

The first US space station to orbit Earth was **Skylab**, launched in 1973. Skylab was manned by three groups of astronauts during 1973 and 1974. The first crew stayed on Skylab for 28 days, the second crew 59 days, while the third and final crew remained aboard for 84 days. Skylab was used to study the long-term effects of space on astronauts and equipment. Skylab fell from orbit and disintegrated in 1979. Debris was scattered from the south-east Indian Ocean into Western Australia.

On 19 February 1986, the Soviet Union launched the 130-tonne space station **Mir**. This station was designed to be permanently manned. However, with the disintegration of the Soviet Union, Mir became a multinational space station. On 8 January 1994, cosmonaut Valeri Polyakov boarded the Mir space station and stayed until March 1995, a record 439 days – long enough to make a trip to Mars. Australian-born astronaut Andy Thomas spent 141 days aboard Mir in 1998 as part of a series of joint missions between Russia and the USA. Mir orbited the Earth 16 times each day for 15 years. During its life, a total of 108 cosmonauts and astronauts lived on Mir. It proved to be an invaluable resource for technological and psychological research for the future International Space Station. Mir finally splashed down in the Pacific Ocean on 23 March 2001.

THE SPACE SHUTTLE

A space shuttle is a reusable spacecraft that is shaped like an aeroplane. It has three main components: the orbiter, the external fuel tank, and the solid booster rockets. The orbiter is a delta-wing vehicle that carries cargo into orbit. The underside of the orbiter is covered by 23 000 ceramic tiles that protect it from the heat generated during re-entry. The external tank fuels and solid boosters fuel the shuttle engines during launch, and are ejected once the fuel runs out. The cargo bay is large enough to hold large satellites. A robotic arm can lift satellites and equipment in and out of the cargo bay. Once in orbit, the shuttle is manoeuvred by means of two small onboard engines. Upon re-entry into the atmosphere, the shuttle follows a shallow glide path and makes a landing without the need of engines. The main advantage of the shuttle over previously used rockets is that it can be reused, and this helps reduce costs.

The world's first reusable manned space vehicle, the space shuttle **Columbia**, was launched on 12 April 1981. Columbia's commander, John Young, was the first astronaut to have flown six missions – he was on the first Gemini mission in 1965. After a flight of over two days, Columbia landed safely back on land at Edwards Air Force base in the USA. Columbia had orbited Earth 37 times. The shuttle was taken back to the Kennedy Space Flight Centre in Florida riding on the back of a Boeing 747 aeroplane.

The USA built five operational space shuttles (Columbia, Atlantis, Discovery, Challenger and Endeavour). It conducted 24 manned flights before tragedy struck on 28 January 1986. On this day the space shuttle

Figure 1.14   The launch of a space shuttle is a spectacular site. (Photo: NASA)

Challenger exploded 73 seconds after being launched, and all seven astronauts on board were killed. The accident forced engineers to re-design various parts of the shuttle, and further flights were delayed until the launch of Discovery on 29 September 1988.

On 1 February 2003, the space shuttle Columbia burnt up on re-entry due to failing heat-shielding tiles. This second shuttle accident resulted in NASA grounding the rest of the fleet. This accident resulted in delays to many NASA projects.

To remain in low Earth orbit, a shuttle must travel at about 28 100 km/h. The exact speed depends on the shuttle's mission and orbital altitude, which normally ranges from 304 km to 528 km. The shuttle normally carries between five and seven crew, but can carry up to eight.

A Russian space shuttle named **Buran** made two unmanned orbits of Earth on 15 November 1988, but the project was later abandoned. The European Space Agency planned to build a shuttle named Hermes, but escalating costs forced the project to be abandoned also.

THE HUBBLE SPACE TELESCOPE

The **Hubble Space Telescope** (HST) was launched from the cargo bay of the shuttle Discovery on 25 April 1990 as a joint venture between NASA and the European Space Agency. The telescope cost $2.5 billion, weighs nearly 12 tonne and orbits 600 km above the Earth at a speed of 28 000 km/h. It consists of a 2.4-metre-diameter mirror mounted in a large tube, three cameras, two spectrographs and a number of guidance sensors. The idea of having a telescope orbiting Earth is to look at objects in the universe without the distorting and obscuring influence of the atmosphere. Hubble can detect objects about a billion times fainter than the human eye. Scientists are using the telescope to learn about the nature of stars, planets and black holes, the evolution of the universe, and distant objects never seen before.

Figure 1.15   The Hubble Space Telescope. (Photo: NASA)

During its lifetime, the HST will study the universe at wavelengths from the infrared through the ultraviolet. Hubble has been used to view over 10 000 objects throughout the universe. Its high-resolution images of Mars, Jupiter, Saturn and Neptune are providing surprising detail about these planets. The world was amazed in July 1994 when Hubble produced images showing the impact of the comet Shoemaker-Levy 9 with Jupiter. In the Great Nebula in Orion, dusty discs visible around protostars are thought to be new solar systems forming. Discs of matter have been seen swirling around super massive black holes at the centre of galaxies and quasars, as well as structure in the spiral arms of nearby galaxies.

Astronomers are already planning the next generation space telescope that will be able to view the universe in even more detail.

DID YOU KNOW?

The Hubble Space Telescope was not an immediate success. The first pictures sent back to Earth were blurred because the telescope's main mirror was not the right shape. Two months after its launch, NASA scientists decided the optics of the telescope needed to be corrected and a space shuttle mission was organised to repair it. In December 1993, astronauts on the shuttle Endeavour worked for six days to repair Hubble so it could see more clearly. The astronauts set new records for spacewalks and the amount of work done while living in space. The corrective optics installed by the astronauts was highly successful and the images are now crystal clear.

OTHER SPACE TELESCOPES

Two other space observatories orbit Earth, Europe's XMM-Newton and America's Chandra X-ray observatories. **Chandra** was launched by the shuttle Columbia in July 1999, and is designed to observe X-rays from high-energy sources, such as the remains of exploded stars. X-rays provide scientists with a different perspective when exploring space. The **XMM-Newton** was launched by rocket in December 1999, and is designed to investigate the origins of the universe by probing cosmic matter from black holes. The elliptical orbits of both observatories takes them from between 7000 and 10 000 kilometres above Earth to more than a third of the way to the Moon.

INTERNATIONAL SPACE STATION

Alpha, the **International Space Station** (ISS), is the largest and most complex international scientific project in history. Construction is a joint

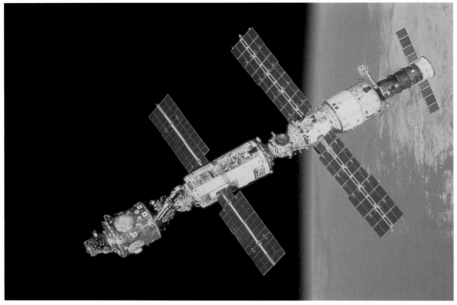

Figure 1.16   The International Space Station. (Photo: NASA)

venture between the USA, Russia, Japan, Canada, Brazil, and 15 countries of the European Space Agency. The ISS is more than four times as large as the Russian Mir space station, and has a mass of over 450 tonnes. There are more than 100 components to be assembled in stages, requiring 46 space flights and over 100 spacewalks. The first two sections of the ISS were assembled in 1998 during a 12-day shuttle mission. In July 2000, Russia's Zvezda service module was connected to the station. When complete, the ISS will measure 108 m across by 88 m long, with almost half a hectare of solar panels powering its six laboratories. It orbits Earth at an altitude of around 400 kilometres.

Teams of up to seven astronauts and scientists will visit the station for periods of up to four months. The ISS will be used to conduct experiments in space that could not be performed on Earth. Research will include studies of protein crystal growth, tissue culture, living in low gravity, behaviour of new materials, space technology and environmental change on Earth. For earth science researchers, the ISS will provide an excellent viewing platform as it crosses over the same area of the planet every three days and covers about 90 per cent of the Earth's surface. A third of the ISS resources will be devoted to commercial enterprises. Many of the research programs planned for the ISS benefit from longer stays in space.

The first long-term crew to live aboard the ISS, commanded by US astronaut Bill Shepherd and including Russian cosmonauts Yuri Gidzenko and Sergei Krikalev, was launched in October 2000 on a Russian Soyuz spacecraft (Expedition 1). In March 2000, Australian-born astronaut Andy Thomas spent a week on the ISS before returning to Earth on the space shuttle Discovery. The longest spacewalk by US astronauts took place when

FACTS ON THE ISS

- The ISS will be larger than a five-bedroom house.
- The ISS will have an internal pressurised volume of about one-and-a-half Boeing 747 aeroplanes.
- Fifty-two computers will control the systems on the ISS.
- The ISS will be about four times larger than the Russian space station Mir, and about five times as large as the US Skylab.
- The electrical system will contain about 13 kilometres of wiring.
- The ISS will manage 20 times as many signals as the space shuttle.
- The 110 kilowatts of power for the ISS will be supplied by almost half a hectare of solar panels.
- The 15-metre-long robotic arm provided by Canada will be able to lift the weight of a space shuttle orbiter.
- The station will orbit at an altitude of 400 kilometres. This orbit allows the station to be reached by launch vehicles of all the international partners.
- It would take the US space shuttle fleet more than a dozen years and 60 flights to achieve what the ISS can achieve in one year in orbit.

the second crew (Expedition 2) arrived at the ISS in March 2001. Jim Voss and Susan Helms worked outside for almost nine hours preparing for the docking of the Italian cargo container, Leonardo. On 28 April 2001, American Dennis Tito became the first space tourist to visit the space station.

When complete the ISS will be the third brightest object in the sky after the Sun and Moon.

THE FUTURE OF SPACE EXPLORATION

The race to explore space advanced rapidly in the 1960s when the first pioneers of spaceflight where placed in Earth orbit. Since then we have seen advances in space technology through the USA's Mercury, Gemini and Apollo programs, and the Soviet Vostok, Voskhod and Soyuz projects. It was largely as a result of these programs that the first human was able to walk on the Moon in 1969. This was perhaps the most significant step in the exploration of the solar system, since the Moon was the first object outside Earth that humans had ventured onto.

Since landing on the Moon, humans have turned their attention to improving methods of space travel and to studying the long-term impact of space travel on humans. As a result of this new focus, space stations like Salyut, Mir, Skylab and ultimately the International Space Station were built. Future manned missions into space will require new technologies and the current space stations are providing a pathway for this to occur.

Table 1.2 Significant Earth–Moon pioneering spacecraft (prior to 2000)

Launched	Name of probe	Launched by	Type
1957	Sputnik 1	USSR	Unmanned
1957	Sputnik 2	USSR	Carried a dog
1958	Explorer 1	USA	Unmanned
1959	Luna 1, 2, 3	USSR	Moon probes
1961	Vostok 1	USSR	Manned – Earth orbit
1961	Freedom 7	USA	Manned – part orbit of Earth
1962	Friendship 7	USA	Manned – Earth orbit
1962	Ranger 4, 5	USA	Moon probe
1963	Vostok 6	USSR	Manned – Earth orbit
1964	Voskhod 1	USSR	Manned – Earth orbit
1965	Gemini 3	USA	Manned – Earth orbit
1965	Voskhod 2	USSR	Manned – Earth orbit
1965	Gemini 4	USA	Manned – Earth orbit
1966	Gemini 8	USA	Manned – Earth orbit
1966	Luna 9	USSR	Moon probe
1967	Soyuz 1	USSR	Manned – Earth orbit
1967	Surveyor 5, 6	USA	Moon probes
1968	Zond 5	USSR	Moon probe
1968	Apollo 7	USA	Manned – Earth orbit
1968	Apollo 8	USA	Manned – Moon fly-by
1969	Apollo 10	USA	Manned – Moon descent
1969	Apollo 11	USA	Manned – Moon landing
1969	Apollo 12	USA	Manned – Moon landing
1970	Luna 16	USSR	Moon probe
1970	Apollo 13	USA	Manned – Moon fly-by
1971	Apollo 14	USA	Manned – Moon landing
1971	Apollo 15	USA	Manned – Moon landing
1971	Salyut 1	USSR	Manned – space station
1972	Apollo 16	USA	Manned – Moon landing
1972	Apollo 17	USA	Manned – Moon landing
1973	Skylab	USA	Manned – space station
1981	Columbia	USA	Space shuttle
1984	Salyut 7	USSR	Manned – space station
1986	Mir	USSR	Manned – space station
1986	Challenger	USA	Space shuttle – exploded
1988	Discovery	USA	Space shuttle
1988	Buran	USSR	Space shuttle – unmanned
1990	Hubble	USA	Space telescope – Earth orbit
1998	International SS	Various	Space station in Earth orbit
1999	Chandra	USA	Space telescope
1999	XMM-Newton	ESA	Space telescope

Web Notes

If you are interested in tracking the movement of the International Space Station from Earth, use the web site <http://spaceflight.nasa.gov/realdata/tracking/>. For regular updates on space launches, missions, and satellites, try <http://www.planet4589.org>.

CHAPTER 2

EXPLORING
THE PLANETS

THIS CHAPTER REVIEWS the significant space probes used to explore the solar system. Many probes were sent into space but were unsuccessful or made insignificant contributions. The probes examined here have added greatly to our knowledge of the planets in the solar system.

EXPLORING SPACE BEYOND EARTH AND THE MOON

Since the early space missions put humans into orbit around Earth, many advances have been made in space technology. It is now possible to send space probes deep into the solar system to explore other planets. Humans have only travelled as far as the Moon and back, but robotic space probes have been placed on the surface of planets like Venus and Mars, as well as on the surface of an asteroid and Saturn's moon Titan. Probes have been used to explore distant planets like Jupiter and Saturn.

Planetary probes orbit around the Sun. They often pass target planets, go into orbit around planets, or land on planets. Instruments on board space probes collect information about the planet and return it to Earth via radio signals. For example, spectroscopes are used to analyse light and

Figure 2.1   Spacewalk by an astronaut outside the space shuttle Atlantis.
(Photo: NASA)

tell what elements are present, infrared detectors are used to measure
temperatures and detect molecular gases, magnetometers measure
magnetic fields, and radar is used to measure surface features.

The first planet to which humans sent a spacecraft was Venus, the closest
planet to Earth. Venus is similar in size and mass to Earth, and has always
been of interest to humans. Mars was the next planet to which a probe was
sent, followed by Mercury. Some of these early space probes were successful
while others failed. All, however, provided information and experience
that led to future success.

Early solar probes

Early spacecraft sent towards the Sun included **Pioneers 5**, **6**, **7**, **8** and **9**.
These probes were launched between 1959 and 1968 by the USA and are
still in solar orbit, but most have ceased functioning.

The US solar probe **Explorer 49**, launched on 10 June 1973, is now in lunar orbit. **Helios 1** and **Helios 2**, launched by the USA and West Germany in 1973 and 1976 respectively, came within about 45 million kilometres of the Sun.

The **Solar Maximum Mission (SMM)** launched in February 1980 by the USA was designed to provide coordinated observations of solar activity, in particular solar flares, during a period of maximum solar activity. The probe malfunctioned in 1981 and was repaired by a space shuttle crew in 1984 and functioned until 1989.

Venera probes

On 12 February 1961, the Soviet Union launched **Venera 1** towards the planet Venus. Two weeks after launch, radio contact with the probe was lost and the probe missed the planet by 97 000 km. It ended up going into orbit around the Sun. Venera 2 also missed the planet, while Venera 3 crashed into Venus.

The first probe to return atmospheric data from the surface of another planet was **Venera 7**, in 1970. A few years later, in 1975, **Veneras 9** and **10** were the first spacecraft to return images of the surface of another planet.

Two more Soviet probes, **Venera 11** and **Venera 12**, were sent to Venus in 1978. Both probes made a chemical analysis of the Venusian atmosphere. In 1981, **Venera 13** sent back coloured photographs of the surface of Venus and analysed soil samples.

Mariner probes

America's first successful probe to fly by another planet (Venus) was **Mariner 2**, in December 1962. Infrared and microwave detectors on Mariner 2 confirmed that Venus was a very hot planet with a cloud-covered surface and an atmosphere composed mainly of carbon dioxide. The probe also discovered that, unlike Earth, Venus did not have a magnetic field or radiation belts. Three weeks after the Venus fly-by, Mariner 2 went off the air and it is now in solar orbit.

Mariner 3 was launched on a mission to Mars on 5 November 1964, but it was lost into interplanetary space when it was unable to collect solar energy for power. The next planetary probe, **Mariner 4**, was successful at reaching Mars in 1965, and it sent back the first close-up images of the Martian surface as it flew within 9846 km of the surface of the planet. The images showed Mars to be a cratered world with an atmosphere much thinner than previously thought. The thin atmosphere was confirmed to be mostly carbon dioxide, and a small magnetic field was detected. Mariner 4 is now in solar orbit. **Mariner 6** (1969) and **Mariner 7** (1969) measured atmospheric conditions and produced 201 photographs, including the first close-up images of the Martian polar cap. The most successful of the early Mars missions was **Mariner 9**, which orbited Mars for 349 days beginning in November 1971. Mariner 9 returned images showing huge volcanoes and giant canyon systems on the red planet's surface, and provided pictorial evidence suggesting that water once flowed across the planet. The probe also took the first close-up views of the two Martian moons, Phobos and Deimos.

The final Mariner probe, **Mariner 10**, was the first dual-planet mission and the first spacecraft to have an imaging system. It flew past Venus on 5 February 1974 for a gravity assist to the planet Mercury. Mariner 10 flew past Mercury three times during 1974 and 1975. The spacecraft returned

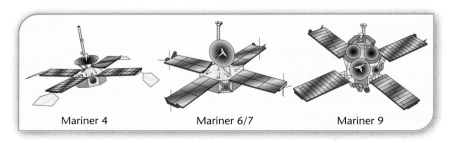

Figure 2.2   Some of the Mariner space probes.

the first close-up images of the atmosphere of Venus in ultraviolet light, and showed that the cloud system circles the planet in four Earth days. It recorded circulation in the Venusian atmosphere and showed the temperature of the cloud tops to be –23°C. Mariner 10 took more than 10 000 photographs of the moon-like cratered surface of Mercury, with 57 per cent planet coverage. A weak magnetic field was detected on Mercury, but no atmosphere. Mariner 10 is now in solar orbit.

The Pioneer probes

The USA began its exploration of more distant parts of the solar system with the Pioneer 10 and 11 space probes, launched in 1972 and 1973. These craft were designed to survive the passage through the asteroid belt and Jupiter's magnetosphere. The asteroid belt was relatively simple to pass through since there were many gaps, but the space probes were nearly fried by ions trapped in Jupiter's magnetic field.

In December 1973, **Pioneer 10** was the first spacecraft to fly by Jupiter, passing within 130 000 kilometres of the cloud-covered surface. Twenty-three low-resolution images were returned to Earth showing Jupiter's turbulent atmosphere and the Great Red Spot. Data were collected on the temperature and pressure in Jupiter's atmosphere, and three of Jupiter's moon's detected – Europa, Ganymede and Callisto. Pioneer 10's greatest achievement was the data collected on Jupiter's magnetic field, trapped charged particles, and solar wind interactions.

About a year later, **Pioneer 11** flew to within 48 000 km of Jupiter's surface and sent back 17 images of the planet. The space probe used the strong gravitational field of Jupiter to swing it on a path towards Saturn, a journey that was to take five years. In September 1979, Pioneer 11 passed within 21 000 km of Saturn's surface, and it returned 440 images and data about Saturn's moons and its ring system.

Today Pioneers 10 and 11 are no longer functioning, but both are still travelling at about 12 km/s and heading in opposite directions away from the solar system into deep space. Each craft carries a plaque with a graphic message to inform anyone out there about the solar system, the Earth, and the human race.

The Viking missions

The Viking space probes were the first spacecraft to conduct long-term research on the surface of another planet. Each of the two unmanned probes consisted of an orbiter and a lander.

The first of the Viking space probes, **Viking 1**, was launched from Cape Canaveral, Florida, on 20 August 1975, aboard a Titan rocket. On 19 June 1976, the probe went into orbit around Mars and a lander from it reached the surface on 20 July. The lander was designed to photograph the surface, collect and analyse soil samples, and take measurements of the atmosphere. The lander showed the surface to consist of red sand and boulders

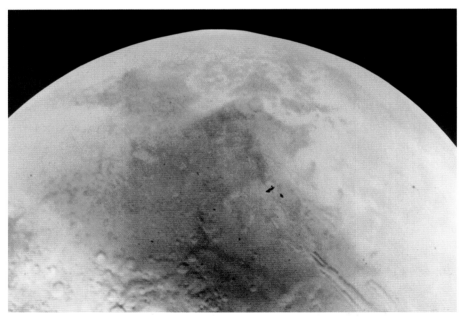

Figure 2.3   The surface of Mars as seen from Viking 1 in June 1976. (Photo: NASA)

stretching out as far as its cameras could see. Although the craft was programmed to search for micro-organisms, no conclusive evidence of life was found. The sky of Mars was found to be pink as a result of sunlight reflecting off the reddish dust particles in the thin atmosphere. Light winds were measured, with gusts up to 50 km/h, and temperatures ranged from −30°C to −85°C.

Viking 2 was launched on 9 September 1975 and arrived at Mars on 7 August 1976. The lander touched down in a different region from Viking 1 on 3 September. This second lander was able to detect a movement in the Martian crust (a Marsquake).

Between them, the landers sent back 4200 photographs from Mars's surface, while the orbiters sent back 52 000 photographs covering most of the planet as seen from orbit. Viking 1 kept functioning until late November 1982, while Viking 2 stopped functioning in April 1980.

The Voyager project

The USA launched **Voyager 1** and **Voyager 2** in 1977 on a mission towards Jupiter, the largest planet in the solar system. These probes flew by Jupiter in March and July of 1979 respectively before proceeding to Saturn. Each Voyager was equipped with high-resolution cameras, three programmable computers, and instruments to conduct a range of scientific experiments. Voyager 2 was launched sixteen days before Voyager 1, but Voyager 1 took a faster, more direct route to reach Jupiter first.

Thousands of photographs were sent back to Earth as the Voyager probes passed Jupiter.

The two probes discovered that Jupiter has a complex, turbulent atmosphere with lightning and auroras. A ring system and three new satellites (moons) were also discovered. Another major surprise was that one of Jupiter's moons, Io, has active sulfurous volcanoes.

The strong gravitational pull of Jupiter was used to fling the Voyager probes towards Saturn. Voyager 1 reached Saturn in November 1980 and

Figure 2.4 In order to track the Voyager spacecraft, NASA uses a system of deep space communications stations scattered across the Earth. One of these is located at Tidbinbilla near Canberra, Australia. (Photo: J. Wilkinson)

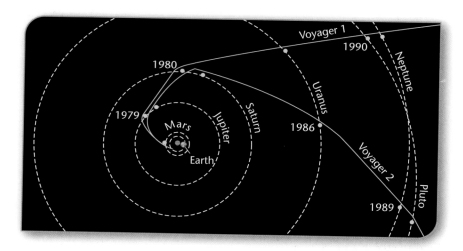

Figure 2.5 Paths of Voyager 1 and Voyager 2 through the solar system. Both probes used the gravitational pull of Jupiter and Saturn to accelerate them further out into the solar system. This process is called gravity assist.

Voyager 2 in August 1981. The probes discovered seven new moons and returned spectacular photographs of the planet's ring system.

After passing Saturn, Voyager 1 left the solar system, but Voyager 2 was able to visit Uranus on January 1986 and Neptune in August 1989. Voyager 2 discovered several new moons around each planet as well as ring systems. Triton, a moon of Neptune, was found to have active geyser-like structures and an atmosphere. After passing over Neptune's north pole, this space probe also headed out of the solar system.

These two space probes provided spectacular close-up views of the four outer planets, known as the gas giants. Both space probes continue to operate beyond the solar system, and are returning data about cosmic rays in outer space and ultraviolet sources among the stars. Communication

DID YOU KNOW?

Halley's comet passed the Earth in 1985–86, and a number of countries sent space probes to observe the comet. The probe that got the closest to the comet was Giotto, launched by the European Space Agency. **Giotto** passed within 600 km of the nucleus of Halley's comet on 13 March 1986. The probe carried 10 instruments including a colour camera. Giotto was severely damaged by high-speed dust encounters as it neared the comet and it was placed into hibernation soon after. In 1990, Giotto was reactivated and some of its instruments still worked. On 2 July 1990, Giotto came close to Earth and was re-targeted to a successful fly-by of comet Grigg-Skjellerup on 10 July 1992.

The US space probes Pioneer Venus, Pioneer 7 and Dynamics Explorer 1 also collected data on Halley's comet. The comet's nucleus was found to be dark in colour, irregularly shaped, about 16 km by 8 km in size, and covered in dust. The comet's inner coma contained a mixture of 80 per cent water vapour, 10 per cent carbon monoxide, and 3.5 per cent carbon dioxide and some organic compounds.

with the probes is expected to continue until 2015, when the radioisotope thermoelectric generators on the probes are expected to run out of power. Voyager 1 passed the Pioneer 10 space probe on 17 February 1998 and is now the most distant human-made object in space.

Magellan

NASA placed the Magellan space probe into orbit via a space shuttle in May 1989. The craft was named after the sixteenth-century Portuguese explorer who first sailed around the Earth. Magellan reached its destination, Venus, in August 1990, with the aim of mapping the surface of Venus using radar. The synthetic aperture radar (SAR) imaging of features on Venus was more than 10 times sharper than previous radar images taken from other orbiting space probes. Other data from Magellan were used to prepare a comprehensive gravity field map for most of the planet, and it executed a first-ever aero-braking manoeuvre, dipping into the atmosphere to change its speed and circularise its orbit. This fuel-saving technique was

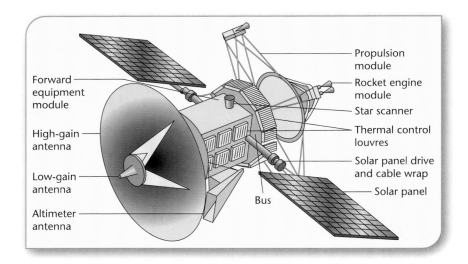

Figure 2.6   The Magellan space probe.

to be used on future missions. Magellan made 15 000 orbits of Venus before contact was lost as the craft descended into the Venusian atmosphere in October 1994, five and a half years after launch.

Galileo

Following the success of the Voyager missions to Jupiter and Saturn, space scientists decided to send another probe to Jupiter in the early 1980s. This probe, named **Galileo** after the Italian astronomer Galileo Galilei (1564–1642), was designed to orbit Jupiter (unlike the Voyager probes which flew by the planet).

The launch of the probe was scheduled for May 1986 from a space shuttle, but the Challenger disaster in January of that year delayed the launch until September 1989. A new three-year flight path had to be planned that used gravity assists from Venus and Earth to sling Galileo towards Jupiter. The flight path took the probe near the asteroids Gaspra in 1991 and Ida in 1993. Cameras on Galileo showed Gaspra might be a rocky fragment of a larger body, while Ida was a rocky body with a small moon-like body orbiting it.

In July 1994 Galileo recorded pictures of the comet Shoemaker-Levy 9 hitting Jupiter.

Galileo finally made it to Jupiter in December 1995. A smaller probe was released from the orbiter and parachuted into the upper cloud layers of Jupiter's atmosphere. The probe returned almost an hour's worth of data before its signal was lost. Scientists found that Jupiter's atmosphere was dry and contained mostly hydrogen. The atmosphere was denser and windier than expected.

Galileo continued to orbit Jupiter until late 2003, collecting and relaying information about and images of Jupiter and its moons back to Earth. It was then deliberately crashed into Jupiter's atmosphere.

Ulysses – a solar probe

Ulysses is a joint NASA and European Space Agency (ESA) mission to study the Sun. The space probe was launched in October 1990 via the space shuttle Discovery. In February 1992, Ulysses received a gravity assist from Jupiter to place it in an orbit that crossed the Sun's poles. The probe has now completed its main mission of surveying the north and south poles of the Sun. As well as gathering data about the solar wind, the probe also collected data about the interplanetary magnetic field, interplanetary dust and cosmic rays.

Ulysses made further passes over the Sun's polar regions in 1994 and 1995 and again in 2000 and 2001. This second set of solar passes took place during a period of high sunspot activity, and allowed Ulysses to measure a wide range of solar activity.

In 2007 and 2008 the European-built space probe flew over the Sun's poles for a third time. The probe will be monitoring changes in the Sun's magnetic field and help to improve 'solar weather forecasting'.

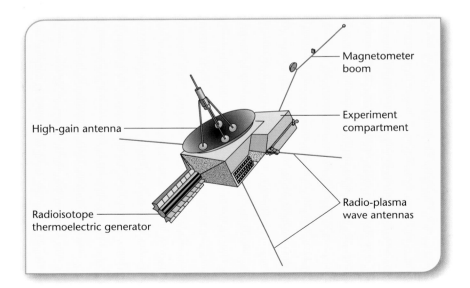

Figure 2.7   The Ulysses space probe.

DID YOU KNOW?

The Ulysses space probe was named after a hero of Greek legend. In the 26th Canto of his poem *Inferno,* the medieval Italian poet Dante describes the last voyage of Ulysses. He has gathered his crew for one final adventure, but storms and setbacks get in the way and his men become restless. Ulysses inspires his wavering comrades with talk of a journey that will search for the uninhabited world behind the Sun, exactly the destination of the ESA–NASA namesake mission.

SOHO

Another joint project of the ESA and NASA, the Solar Heliospheric Observatory, or **SOHO**, was launched in December 1995 via an Atlas Centaur rocket. SOHO has been keeping a watch on the Sun since April 1996 and is in orbit around the Sun in a region between the Earth and Sun where the gravitational pull of the two bodies are in balance (about 1.5 million km from Earth toward the Sun).

SOHO contains instruments that are studying changes in the Sun's interior, corona, and solar wind. One of SOHO's instruments, called LASCO (Large Angle and Spectrometric Coronagraph), monitors a huge region of space around the Sun 24 hours a day. Two other instruments, SWAN and MDI, allow scientists to 'see' what is happening on the far side of the Sun.

SOHO finished its planned two-year study of the Sun's atmosphere and surface in April 1998. Communications with the spacecraft were interrupted for a while in June 1998, but regained in September of that year. It is still operating 10 years later.

DID YOU KNOW?

Not all interplanetary space probe missions are successful. The first three of the USSR's Mars probes in 1960 and 1962 failed to reach Earth orbit. Mariner 8, launched by the USA in 1971, also failed to reach Earth orbit. The Mars Climate Orbiter, launched by NASA in December 1988 and designed to function as an interplanetary weather satellite, was lost on arrival at the planet. Mars Observer, launched in September 1992, was also lost shortly before its arrival at Mars. An ambitious mission to set another space probe, known as the Mars Polar Lander, down on the frigid terrain near the edge of Mars's south polar cap also ended in failure during descent and landing.

NEAR – mission to an asteroid

On 17 February 1996, NASA launched the **NEAR** (Near Earth Asteroid Rendezvous) space probe via a Delta 2 rocket. The probe's main destination was the asteroid known as 433 Eros. Eros is one of the largest and best-observed of the asteroids whose orbit crosses that of Earth. NEAR began orbiting Eros in February 2000. The probe surveyed the asteroid at altitudes as close as 24 km and sent back images of the surface. Although NEAR was never designed to land on the asteroid, it did make a successful landing on 12 February 2001. Thrusters on NEAR were used to slow its descent to about 8 km/h prior to impact. NEAR's safe landing was confirmed when mission control received a signal from the craft resting on the surface of Eros, some 320 million km from Earth.

Before hitting the surface, NEAR took many close-up pictures of the asteroid. The pictures revealed craters, mysterious bright spots, and boulders the size of soccer fields.

Eros became only the fifth body in the Solar System touched by a human spacecraft, following the Moon, Mars, Venus and Jupiter. Communications with the NEAR spacecraft ended in 2001.

Mars Global Surveyor

Mars Global Surveyor is the first in a series of spacecraft to explore the red planet Mars. NASA launched the first probe in the series on 7 November 1996, and it entered orbit around Mars in September 1997. The probe orbits Mars once every two hours in a nearly circular path that passes over the Martian poles at an altitude of around 400 km.

Scientific instruments on the spacecraft included a thermal emission spectrometer (to examine minerals), a laser altimeter (to produce the first three-dimensional images of the surface topology), a magnetometer (to measure the magnetic field), a camera, and radio relay system.

The mission is providing a new global map of the Martian surface so that scientists on Earth can better understand the evolution of Mars. Images of the surface have shown eroded gullies and other geological features that suggest water once flowed over the surface. The probe is also examining the ionosphere of Mars (a region of charged particles in the

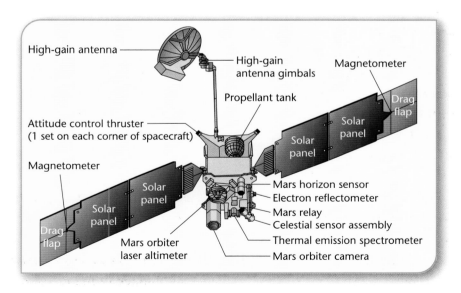

Figure 2.8   The Mars Global Surveyor.

atmosphere), weather patterns, water vapour distributions and the geology of the interior of Mars.

Mars Pathfinder

Mars Pathfinder is the second probe of the Mars Surveyor Program conducted by NASA. The mission consists of a stationary lander and surface rover (robot) known as Sojourner. The mission's main objective was the exploration of the surface of Mars at a low cost.

The probe was launched in December 1996 and landed successfully on Mars in July 1997 following a descent using parachutes, rockets and airbags to soften the landing.

The landing site, an ancient floodplain in Mars's northern hemisphere known as Ares Vallis, is among the rockiest parts of Mars. Once on the ground, three solar panels were unfolded to provide power for the lander.

Figure 2.9   The Mars Surveyor lander. (Photo: NASA)

Scientific instruments carried aboard Pathfinder included the lander's camera, the atmospheric structure instrument/meteorology package, and the rover's alpha proton X-ray spectrometer. The lander's camera was a stereo imaging system with navigational and colour capability. The atmospheric and meteorology package acquired atmospheric data during the descent of the lander as well as from the ground. The alpha proton X-ray spectrometer was designed to determine the elements that make up the rocks and soil on Mars.

After landing, Mars Pathfinder sent back to Earth more than 16 000 images from the lander and 550 images from the rover, as well as more than 15 chemical analyses of rocks and extensive data on winds and weather factors.

The Pathfinder mission lasted three times longer than expected and returned a large amount of data. The mission ended late in 1997 when communication with the probes was lost because the main batteries on the spacecraft and rover ran out of power.

Major milestones of the mission were the ability to place a spacecraft on the surface of another planet by way of direct entry into its atmosphere, and the ability to explore the surface of a planet using a roving robotic machine.

DID YOU KNOW?

The Pathfinder rover was named after Sojourner Truth, an African-American campaigner who lived during the time of the US Civil War and travelled throughout the country advocating the rights of all people to be free. The name Sojourner, which means 'traveller', was the winning entry in a year-long competition in which students 18 years and younger submitted essays on the historical accomplishments of a heroine of their choosing.

Cassini

Launched on 15 October 1997, the **Cassini** mission to Saturn is a joint project of NASA, the ESA and the Italian Space Agency. The spacecraft, which included the Huygens probe, the launch adaptor and the propellant, was about the size of a school bus and weighed 5.6 tonnes, making it the third heaviest craft launched into space. The probe cost $4.7 billion and contained a record number of 12 instruments to help it explore Saturn's rings, its 31 known moons, and magnetosphere. Cassini was also equipped with special radio antennae to record sounds.

Flying directly to Saturn would have required much more fuel than Cassini could carry, so gravity assist fly-bys of Venus, Earth and Jupiter were used. The craft reached Saturn in June 2004. In its first week around Saturn, Cassini detected the moaning sound generated by the solar wind hitting Saturn's magnetic field.

Cassini flew between Saturn's two outer rings at about 80 000 km/h before it slowed down enough to be captured by Saturn's gravity and begin orbiting the planet.

The Huygens probe on board Cassini descended to the surface of Titan, Saturn's largest moon, in January 2005, sending back data and images. The Cassini orbiter used its onboard radar to peer through Titan's clouds and determine if there is liquid on the surface. Experiments aboard both orbiter and the entry probe investigated the chemical processes that produce this unique atmosphere. Titan is of interest to scientists because its atmosphere contains chemical compounds similar to those that resulted in life on Earth.

Mars Odyssey

Mars Odyssey is an orbiting space probe designed to determine the composition of the Martian surface, to detect water and shallow buried ice, and to study environmental radiation. The craft was launched on

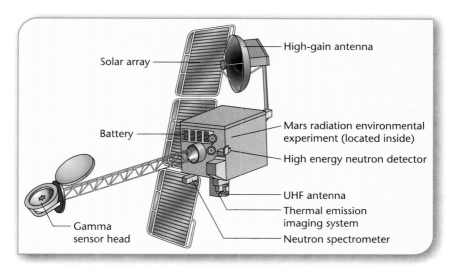

Figure 2.10   The Mars Odyssey space probe.

7 April 2001 via a Delta 2 launch vehicle from Cape Canaveral, Florida, and arrived at Mars in October of that year.

The probe began orbiting the planet once every 25 hours, but this period was shortened to 2 hours as the craft's orbital path decreased bringing it closer and closer to Mars.

One of the aims of the Odyssey mission was to find out more about the geology of Mars. A gamma-ray spectrometer was used to map the surface for elements like hydrogen, silicon, iron, potassium, thorium and chlorine. These and other chemical elements are the building blocks of minerals, and minerals are the building blocks of rocks. Rocks and land formations tell scientists much about the past and present history of Mars. The spectrometer is also able to determine the thickness of the polar ice-cap.

Infrared cameras on Odyssey could distinguish between sandy areas and rocky areas beneath the surface.

The radiation experiment was designed to collect data on the radiation environment in space near Mars to help assess potential risks to any future human space travellers.

Genesis Mission

NASA launched the Genesis space probe in August 2001. The main objective of the mission was to collect samples of solar wind particles and return them to Earth for detailed analysis. Scientists wanted to measure oxygen and nitrogen isotopes in order to determine the role of each type of isotope in the formation of the solar system.

Genesis was the first return to Earth by a spacecraft containing samples since the US Apollo and Soviet Luna missions brought back moon rocks in the 1970s.

Mars Exploration Rovers

In 2003 NASA launched two **Mars Exploration Rovers** with the purpose of exploring the surface of Mars. The first probe was launched on 10 June 2003, the second on 7 July 2003; and both landed on Mars in January 2004. During the trip to Mars, the first probe made four trajectory correction manoeuvres and the second probe performed three. The two probes survived blasts of high-energy particles from some of the most intense solar flares on record. Each probe was used successfully to place a rover (mobile robot) on the Martian surface, but on opposite sides of the planet. One rover was called Opportunity, the other Spirit. Each rover was parachuted to the surface. Air bags were used to cushion the impact. The rovers were designed to look for geological clues about Mars, including whether parts of Mars formerly had environments wet enough to be hospitable to life. Each rover has the capability to explore its surroundings for interesting rocks and soils, to move to those targets, and to examine their composition and structure.

In late September 2004, the mission of both rovers was extended. The solar-powered machines were still in good health after two weeks of non-use while communications were unreliable because Mars was passing behind the Sun. As of 1 January 2008, the rovers are still operating and are still

Table 2.1 Significant missions to explore the solar system

Launched	Space probe	Launched by	Arrival	Destination
1961	Venera 1, 2, 3	USSR	1961	Venus
1962	Mariner 2	USA	1963	Venus
1964	Mariner 4	USA	1965	Mars
1970	Venera 7	USSR	1970	Venus
1971	Mariner 9	USA	1971	Mars
1971	Mars 2	USSR	1971	Mars
1971	Mars 3	USSR	1971	Mars
1972	Pioneer 10	USA	1973	Jupiter
1973	Pioneer 11	USA	1974	Jupiter, Saturn
1973	Mars 4	USSR	1974	Mars
1973	Mars 5	USSR	1974	Mars
1973	Mars 6	USSR	1974	Mars
1973	Mars 7	USSR	1974	Mars
1973	Mariner 10	USA	1974	Venus, Mercury
1975	Venera 11, 12	USSR	1975	Venus
1975	Viking 1	USA	1976	Mars
1975	Viking 2	USA	1976	Mars
1977	Voyager 1	USA	1979	Jupiter, Saturn
1977	Voyager 2	USA	1979	Jupiter, Saturn
1981	Venera 13	USSR	1981	Venus
1985	Giotto	USA	1986	Halley's Comet
1989	Magellan	USA	1990	Venus
1989	Galileo	USA	1995	Jupiter
1990	Ulysses	USA, ESA	1994	Sun
1995	SOHO	USA, ESA	1996	Sun
1996	NEAR	USA	2000	Eros (asteroid)
1996	Mars Global Surveyor	USA	1997	Mars
1996	Mars Pathfinder	USA	1997	Mars
1997	Cassini	USA, ESA, Italy	2004	Saturn
2001	Mars Odyssey	USA	2001	Mars
2001	Genesis	USA	2004	Solar wind
2003	Mars Exploration Rovers	USA	2004	Mars
2003	Mars Express	ESA	2005	Mars
2004	Messenger	USA	2011	Mercury
2005	Mars Reconnaissance	USA	2006	Mars
2005	Venus Express	ESA	2006	Venus
2006	New Horizons	USA	2015	Pluto
2007	Phoenix	USA	2008	Mars

exploring the surface. It is amazing that the two craft have survived for so long.

Recent comet encounters

Comets can tell astronomers much about the history of the solar system. Only a few space probes have been used to explore the nature of comets.

NASA launched the Deep Impact space probe on 12 January 2005. The probe intercepted comet Tempel 1 in July 2005 and passed within 500 km of its nucleus. The probe fired a 370 kg copper slug into the surface of the comet at 10 km/s. Although Deep Impact collected vast amounts of data on the comet and the collision, it got no clear picture of the resulting crater; it was gone by the time the dust and vapour cleared. Tempel 1, a famous short period comet (5.5 years), is now out at its aphelion 4.7 AU from the Sun, just within the orbit of Jupiter.

On 5 December 2008 the Deep Impact space probe will pass within 700 km of the comet Boethin. Two colour cameras and an infrared spectrometer will be used to observe the comet. Along the way, Deep Impact will act as a space telescope, making photometric observations of several stars with transiting exoplanets to improve our knowledge of their properties.

WEB NOTES

For additional information about spacecraft missions, try:
    <http://solarsystem.nasa.gov>
    <http://www.nineplanets.org/spacecraft.html>
    <http://pds.jpl.nasa.gov>
The following site contains an Astronomy Picture of the Day:
    <http://apod.nasa.gov/apod/>. Click on the index for links to solar system and
    space technology.

CHAPTER 3

THE SUN

THE SUN IS the dominant object in the solar system because it is the largest object. It is positioned at the centre of the solar system and its gravitational pull holds all the planets in orbit. The Sun is an average-sized **star** about 4.5 billion years old. Unlike planets, stars produce their own light and heat by 'burning' fuels like hydrogen and helium in a process known as **nuclear fusion**. Stars have a limited life and the Sun is no exception – it is about half way through its lifespan of about 10 billion years.

The Sun is one of over 100 billion stars that make up the **galaxy** called the **Milky Way**. The Milky Way galaxy is spiral in shape, and the Sun is positioned in the plane of the galaxy about half-way out from the centre. The Milky Way is about 100 000 **light years** in diameter and 15 000 light years thick. You can see parts of the Milky Way as a band or cloud of stars that stretches across the night sky. Within the Milky Way, the Sun is moving at 210 km/s, and takes 225 million years to complete one revolution of the galaxy's central mass of stars.

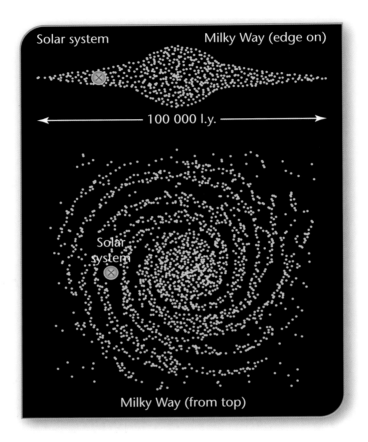

Figure 3.1   Position of the Sun in the Milky Way galaxy.

PROBING THE SUN

Scientists have gained much of their knowledge about the Sun from many years of observations made from Earth. However, much of our more recent knowledge has come from space probes that have been sent on missions to the Sun. These probes have provided accurate information about the Sun's temperature, atmosphere, composition, magnetic fields, flares, prominences, sunspots and internal dynamics.

DID YOU KNOW?

Astronomers use light years as a measure of distance in the universe. One light year is the distance that light travels in one Earth year. The speed of light is 300 000 kilometres per second, so in one year light travels a distance of 9.5 million million kilometres. The nearest star to our solar system (apart from the Sun) is Proxima Centauri, at a distance of about 4.2 light years. In a scale model with the Sun and Earth 30 cm apart, Proxima Centauri would be 82 km away. Because Proxima Centauri is 4.2 light years away from Earth, it takes 4.2 years for light from that star to reach us. Thus when you look up into the night sky at this star, the light you see left the star 4.2 years ago. You are really looking into the past.

It takes about 8 minutes 20 seconds for light to travel from the Sun to Earth.

Table 3.1 Details of the Sun

Mass	2.0×10^{30} kg
Size relative to Earth	109 times larger
Diameter	1.4 million km
Distance from Earth	150 million km
Density	1410 kg/m³
Luminosity	3.9×10^{26} J/s
Surface temperature	5500°C
Interior temperature	15 million°C
Equatorial rotation period	25 days
Composition	92% hydrogen, 7.8% helium
Surface gravity	290 N/kg (29 × Earth gravity)
Escape velocity	618 km/s
Photosphere thickness	400 km
Chromosphere thickness	2500 km
Core pressure	250 billion atmospheres
Sunspot cycle	11 years
Age	4.5 billion years

The USA launched a number of unmanned solar probes between 1959 and 1968 as part of its **Pioneer** program. Many of these early probes have now completed their missions but still remain in orbit around the Sun. Missions such as Pioneers 10 and 11 showed that gravity assists were possible and that spacecraft could survive in high-radiation areas.

America's first space station, **Skylab** (launched in 1973), was used to study the Sun from Earth orbit. The space station included the Apollo Telescope Mount (ATM), which astronauts used to take more than 150 000 images of the Sun. Skylab was abandoned in February 1974 and re-entered the Earth's atmosphere in 1979.

Significant solar probes launched as joint USA and ESA projects include Ulysses (launched 6 October 1990) and SOHO (launched December 1995).

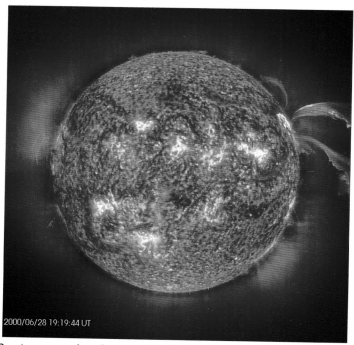

2000/06/28 19:19:44 UT

Figure 3.2   A spectacular photograph of solar flares and the Sun's surface as seen by the SOHO space probe in June 2000. Some of these flares extend more than 500 000 km into space. (Photo: NASA)

In 1998, NASA launched the **TRACE** (Transition Region and Coronal Explorer) satellite into Earth orbit to observe the Sun and study the connection between the Sun's magnetic field and the heating of its corona. The probe was launched from a Pegasus XL rocket dropped from a jet aeroplane flying high above the Pacific Ocean. TRACE's main instrument is a Cassegrain telescope, 30 cm in diameter and 160 cm long. TRACE is one of several small satellites in NASA's Small Explorer (SMEX) project.

The **Genesis** space probe launched by NASA in August 2001 collected samples of solar wind particles and returned them to Earth for analysis. Initial tests showed Genesis collected about 0.4 milligrams of solar particles, equal only to a few grains of salt. They are still being analysed.

Table 3.2 Significant probes sent to the Sun

Name of probe	Country of origin	Launched	Comments
Pioneer 5	USA	1959	
Pioneer 6	USA	1965	Still functioning but no further tracking after December 2000
Pioneer 7	USA	1966	Last contacted 1995
Pioneer 8	USA	1967	Last tracked successfully on 22 August 1996
Pioneer 9	USA	1968	Still in orbit, but failed in 1993
Skylab	USA	1973–79	Space station in Earth orbit
Explorer 49	USA	1973	
Helios 1	USA, Germany	1974–75	Came to within 47 million km of the Sun
Solar Maximum Mission	USA	1980–89	Monitored solar flares
Yohkoh	Japan, USA, UK	1991	Studied high energy radiation from Sun
Helios 2	USA, Germany	1976	Came to within 43 million km of the Sun
Ulysses	USA, ESA	1990	Orbited Sun's polar regions
SOHO	USA, ESA	1995	Study solar wind, corona, internal structure
TRACE	USA	1998	Study solar magnetic fields, corona
Genesis	USA	2001	Collected solar wind particles and returned them to Earth

COMPOSITION OF THE SUN

The Sun is a huge ball of gas that contains about 99 per cent of the mass of the whole solar system. This makes it over 300 000 times as massive as the Earth. The Sun's diameter of 1.4 million km far exceeds Earth's diameter of only 12 760 km. Even the biggest planet – Jupiter – is only one-tenth the diameter of the Sun.

The main elements present in the Sun are hydrogen (92 per cent), followed by helium (7.8 per cent), and less than 1 per cent of heavier elements like oxygen, carbon, nitrogen and neon. The Sun is entirely gaseous, with an average density 1.4 times that of water. Because the pressure in the core is much greater than at the surface, the core density is eight times that of gold, and the pressure is 250 billion times that on Earth's surface.

ENERGY AND LUMINOSITY

The Sun produces a hundred million times more energy than all the planets combined. Just over half this energy is in the form of visible light, with the rest being infrared (heat) radiation. Only about a billionth of the Sun's energy reaches us here on Earth.

The Sun's energy comes from the 'burning' of its hydrogen gas via the process of nuclear fusion. In this process four hydrogen atoms combine to make one helium nucleus. During this process some mass is lost, and it is this mass that is converted into energy. Every second the Sun converts over 600 million tonnes of hydrogen into helium, and this results in 4.5 million tonnes of mass being converted into energy every second.

Energy generated in the core is carried outward to the surface by radiation and convection processes. Core temperature is about 15 million degrees Celsius, while at the surface the temperature is around 5500 degrees Celsius. The surface and interior temperatures are too high for the Sun to have any liquid or solid material.

The **luminosity** of a star is an indication of the total amount of energy it produces every second. This rate depends on the core temperature and pressure of the star, which in turn depends on its mass and age. The Sun's luminosity is 3.9×10^{26} joules per second.

Throughout its life the Sun has increased its luminosity by about 40 per cent and it will continue to increase for some time.

Features of the Sun

The Sun has several different layers. At the centre is the core, which is where energy is produced via nuclear fusion reactions. Above this is the radiative zone, where energy travels very slowly upwards. Closer to the surface is the convective zone, where heat is transported much faster to the surface, or **photosphere**. The photosphere is a thin shell of gases about 200 km thick and forms the visible surface of the Sun. Most of the energy radiated by the Sun passes through this layer. It has a temperature of about

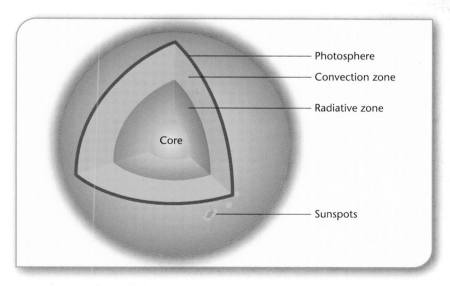

Figure 3.3   The internal structure of the Sun.

5500°C. From Earth the surface looks smooth, but it is actually turbulent and granular because of convection currents. Material boiled off from the surface of the Sun is carried outward by the **solar wind**.

The surface of the Sun also contains dark areas called **sunspots**. Sunspots appear dark because they are cooler than the surrounding photosphere – about 3500°C compared with 5500°C. They radiate only about one-fifth as much energy as the rest of the photosphere.

Sunspots vary in size from 1000 km to over 40 000 km across. As they move across the surface of the Sun, sunspots usually change shape – some disappear, and new ones appear. Their lifetime seems to depend on their size, with small spots lasting only several hours while larger spots may persist for weeks or months. The rate of movement of sunspots can be used to estimate the rotational period of the Sun. At the equator, sunspots take about 25 days to move once around the Sun. At the poles, sunspots take about 36 days to go around the Sun. This odd behaviour is possible because the Sun is not a solid body like the Earth. Sometimes sunspots appear in isolation, but often they arise in groups.

Sunspots and sunspot groups are directly linked to the Sun's intense magnetic fields. Such spots are areas where concentrated magnetic fields break through the hot gases of the photosphere. These magnetic fields are so strong that convective motion beneath the spots is greatly reduced. This in turn reduces the amount of heat brought to the surface compared with the surrounding area, so the spot becomes cooler. Data obtained from space probes like SOHO have shown that the strength of the magnetic fields around sunspots is thousands of times stronger than the Earth's magnetic field. Sometimes these magnetic fields change rapidly, releasing huge amounts of energy that hurl out massive amount of gas as **solar flares** into the Sun's atmosphere.

A typical sunspot is about 10 000 km across. Each has two parts: a black central region called the umbra, which in turn is surrounded by a greyer region, the penumbra. The darker the area the lower the temperature. It is possible to view sunspots from Earth by projecting the image of the Sun

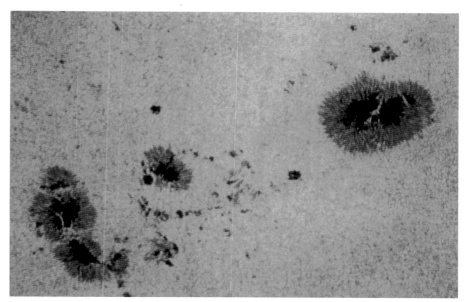

Figure 3.4   Sunspots on the surface of the Sun. Large sunspots contain dark umbral centres, gray penumbral haloes, many large and small single and overlapping spots, and surrounding whitish plages. (Photo: J. Wilkinson)

from a telescope onto a white screen. Observations of the Sun in this way need to be made carefully so as not to damage the viewer's eyes – NEVER look at the Sun through a telescope!

Like many other features of the Sun, the number and location of sunspots vary in a cycle of about 11 years. Heinrich Schwabe, a German astronomer, first noted this cycle in 1843. Sunspot maximums occurred in 1968, 1979, 1990 and 2001. Sunspot minimums occurred in 1965, 1976, 1986, 1997 and 2008. The average latitude of sunspots also varies throughout the sunspot cycle. At the beginning of a sunspot cycle, most sunspots are at moderate latitudes, around 28° north or south. Sunspots arising much later in each cycle typically form closer to the Sun's equator. The variation in the number of sunspots is now known to be the most visible aspect of a profound oscillation of the Sun's magnetic field that affects other aspects of both the surface and interior.

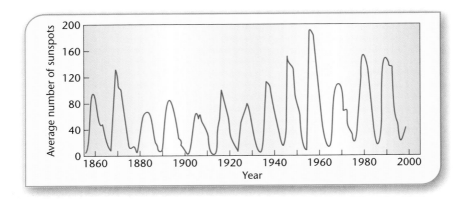

Figure 3.5 Variations in sunspot numbers tend to go through a maximum–minimum cycle every 11 years.

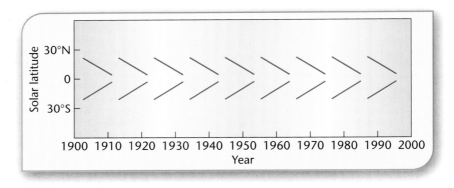

Figure 3.6 Variations in the average latitude of sunspots.

Ejection of material from the surface of the Sun often follows solar flares or other solar phenomena. Sometimes this material reaches Earth and gets trapped in the magnetic field around Earth's polar regions. This material consists mostly of charged particles (ions and electrons), which interfere with communication systems and produce magnetic and ionospheric disturbances such as auroras.

Another visible phenomenon of the photosphere associated with the Sun's magnetic field is the faculae. These are irregular patches or streaks brighter than the surrounding surface. They are clouds of incandescent gas in the upper regions of the photosphere. Such clouds often precede the appearance of sunspots.

The **chromosphere** is the first layer of the Sun's atmosphere. It lies just above the photosphere and is a few thousand kilometres thick. During a solar eclipse, when the Moon passes in front of the Sun, the chromosphere appears as a red shell around the Sun. The chromosphere is much hotter than the photosphere, ranging from 4200°C near the surface to 8200°C higher up. It consists largely of hydrogen, helium and calcium.

The **corona** is the upper layer of the Sun's atmosphere. During a solar eclipse, it appears as a pale white glowing area around the Sun. Temperatures in the corona reach as high as one million degrees Celsius because of interactions between gases and the photosphere's strong magnetic fields. The corona can extend millions of kilometres into space. The corona consists mainly of ionised gas or plasma.

When viewed with a special filter called a hydrogen-alpha filter, dark features called filaments appear in the corona. These are huge masses of

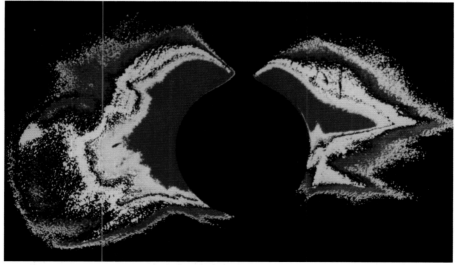

Figure 3.7   A colour-enhanced image of the solar corona, taken by Skylab. Different colours represent different densities of the coronal plasma. (Photo: NASA)

Figure 3.8   Solar prominences. (Photo: J. Wilkinson)

burning gas ejected upwards from the photosphere and suspended in the Sun's magnetic field. When viewed side on (such as at the edge of the Sun), these eruptions, called **prominences,** can be seen as gigantic loops or arches. Temperatures in the prominences can reach 50 000°C. Some prominences last for only a few hours, while others last for weeks. Prominences can only be seen in hydrogen-alpha light using special telescopes or during a total eclipse.

Solar flares occur when the magnetic field of the Sun changes rapidly to create an explosion of charged particles through the Sun's corona. Such events last from a few minutes to a few hours and can send charged particles, X-rays, ultraviolet rays and radio waves into space. Flares can release energy equivalent to more than a billion one-megaton thermonuclear explosions in a few seconds. They are sometimes so violent that they cause additional ionisation in the Earth's ionosphere and may disrupt radio communications.

The solar wind is an erratic flow of highly ionised gas particles that are ejected into space from the Sun's upper atmosphere. The wind produces a huge bubble in space called the **heliosphere**. This wind has large effects on the tails of comets and even has measurable effects on the trajectories of spacecraft.

The Ulysses space probe is providing the first-ever three-dimensional map of the heliosphere from the equator to the poles. Instruments on board the Ulysses space probe have found that the solar wind blows faster around the Sun's poles (750 km/s) than in equatorial regions (350 km/s).

The SOHO space probe found that the solar wind originated from honeycomb-shaped magnetic fields surrounding large bubbling cells near the Sun's poles.

Near Earth, the particles in the solar wind move at speeds of about 400 kilometres per second. These particles often get trapped in Earth's magnetic field, especially around the poles, and produce auroras.

A coronal hole is a large region in the corona that is less dense and is cooler than its surrounds. Such holes may appear at any time during a solar cycle but they are most common during the declining phase of the cycle. Coronal holes allow denser and faster 'gusts' of the solar wind to escape the Sun. They are sources of many disturbances in Earth's ionosphere and magnetic field.

While most of the Sun's activity follows the 11-year sunspot cycle, conditions in the heliosphere are driven by a 22-year magnetic cycle. The Sun's magnetic field is like that of a giant bar magnet with a north and south pole. Data from the Ulysses space probe showed that during the 2001 solar maximum, the Sun's north and south poles changed places. Ulysses next passed over the Sun's poles during the period of the 2007–08 solar minimum. At this time the Sun's magnetic polarity was opposite to that of the previous solar minimum. The magnetic field of the Sun influences the way in which charged particles move through the heliosphere.

SOLAR ECLIPSES

A total solar **eclipse** is one of the most spectacular astronomical events that can be seen from Earth.

Such an event occurs when the Moon passes directly between the Sun and Earth. An eclipse does not occur at every new moon because the Moon's orbit often passes above or below the Sun instead of directly across it. During an eclipse, the Moon's shadow traces a curved path across the surface of the Earth. Any person standing in the path of the shadow will see the sky and landscape go dark as the Moon blocks out the sunlight.

Solar eclipses can be total, partial, or annular, depending on how much of the Sun is covered by the Moon. Total eclipses occur when the Moon is exactly in line between the Earth and Sun and exactly covers the disc of the Sun. If the Moon is not exactly in line between the Earth and Sun, a partial eclipse occurs. An annular eclipse occurs when the Moon is far enough away from Earth that its apparent size is smaller than the Sun's. Hence a bright ring (annulus) remains visible, and the Sun's corona cannot be seen.

There are as many as two total solar eclipses a year, and sometimes as many as five, but few people have a chance to see them. The paths along

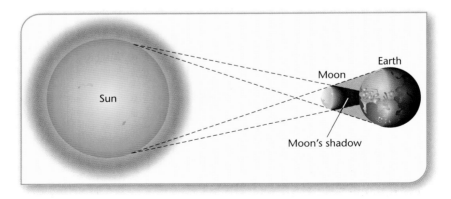

Figure 3.9   How an eclipse of the Sun occurs.

Figure 3.10 The solar corona as seen during a total solar eclipse.
(Photo: J. Wilkinson)

which eclipses can be seen are narrow, and the darkest period, or totality,
can last only about seven and a half minutes at most.

Among the features of a total eclipse are the so-called Bailey's Beads.
These are seen just as the Moon's black disc covers the last thin crescent of
the Sun. Sunlight shining between the mountains at the Moon's edge looks
like sparkling beads. The Diamond Ring effect is a fleeting flash of light
seen immediately preceding or following totality.

At the time of totality, an observer with a small telescope can see the
Sun's prominences as long flame-like tongues of incandescent gases
around the edge of the Moon's disc. Also during totality, the corona can be
seen as a region of glowing gases stretching out from the blacked-out Sun.

Care must be taken when observing the Sun, even during an eclipse – and advice should be sought to protect your eyes from damage.

Future total solar eclipses are due on 22 July 2009 (Asia); 11 July 2010 (South Pacific ocean).

THE SUN'S FUTURE

The Sun is about 4.5 billion years old and it is not quite half-way through its lifespan. Throughout the second half of its life, the Sun is expected to increase gradually in size, luminosity and temperature. In about 5 billion years' time, the Sun will have expanded to about three times its present size. As it uses up its hydrogen it will become more orange in colour. By this time, temperatures on Earth will be much higher and all the water will have evaporated. As the Sun continues to expand, reaching about

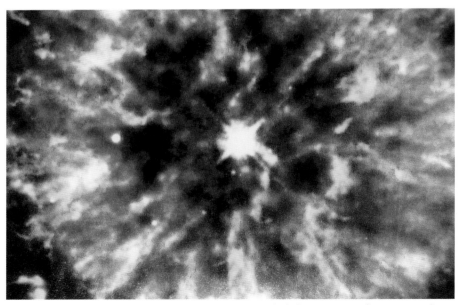

Figure 3.11 The Hubble Space Telescope recorded this picture of a star, similar to our Sun, exploding and sending out much of its material into space. (Photo: NASA)

100 times its present size, it will become a red giant and engulf Mercury and Venus. Earth will be scorched to a cinder. As hydrogen is used up, the core of the Sun will slowly contract, forcing the Sun's central temperature to increase. When the core temperature reaches 100 million degrees Celsius, helium fusion begins to generate carbon and oxygen. The temperature in the core continues to rise, causing helium to fuse at an increasing rate. An explosion (the helium flash) will result, and a third of the Sun will be blown away. Eventually the Sun will lose its outer layers and contract to become a **white dwarf** about the size of Earth.

WEB NOTES

Further information about the Sun and solar space probes can be found at:
    <http://solarsystem.nasa.gov/missions>
    <http://www.esa.int/science/ulysses>
    <http://www.nasa.gov/missions/index.html>
    <http://www.nineplanets.org/sol.html>
    <http://sohowww.nascom.nasa.gov/>

CHAPTER 4

MERCURY – A HOT, OLD WORLD

MERCURY IS THE planet in the solar system nearest to the Sun, with an average orbital radius of 58 million kilometres. It is also the smallest planet of the inner solar system, with a diameter of only 4880 km, making it about the size of our Moon. Mercury travels fast, taking just 88 days to orbit the Sun, but it takes a slow 58.65 days to rotate once on its axis. Of all the **terrestrial planets**, Mercury's orbit is the most elliptical. Its elliptical orbit and slow rotation give it large variations in surface temperatures. During the day temperatures can reach a blistering 430°C, while at night they can drop to a freezing –180°C. No other planet experiences such a wide range of temperatures.

Mercury is thought to have formed at the same time as the other planets in the solar system, about 4.5 billion years ago. Because it is so close to the Sun, Mercury must have been very hot and in a molten state before it cooled to become a solid planet. As Mercury cooled, it also began to contract. The surface of Mercury has been churned up by many meteorite impacts.

EARLY VIEWS ABOUT MERCURY

Mercury has been known since the time of the Sumerians (3rd millennium BC). The planet was given two names by the ancient Greeks: Apollo, for its apparition as a morning star, and Hermes as an evening star. Greek astronomers knew, however, that the two names referred to the same body. To the Greeks, Hermes was the messenger of the gods. In Roman mythology, Mercury was the god of commerce, travel and thievery. The planet probably received this name because it moves so quickly across the sky.

The first map of Mercury was made in the 1880s by Giovanni Schiaparelli, using a simple telescope. The map showed only areas of dark and light. A more detailed map was produced by Eugenios Antoniadi between 1924 and 1933, but has since been proved inaccurate. Both these astronomers believed Mercury rotated once on its axis in 88 Earth days, with one hemisphere permanently facing the Sun. This meant that it was thought that Mercury's day was the same length as its year. However, radar measurements carried out in the early 1960s showed that the true axial rotation period was 58.65 days. Thus it is now known that Mercury rotates three times during two orbits of the Sun. The result of this is that the same hemisphere is pointed towards Earth every time the planet is best placed for observation. This effect also means that the Mercurian day (sunrise to sunset) is 176 Earth-days long, or two Mercurian years.

Antoniadi also believed that Mercury had an atmosphere because he thought he

Figure 4.1 Mercury as seen by the Mariner 10 space probe in 1974 from 200 000 km away. (Photo: NASA)

74

Table 4.1 Details of Mercury

Distance from Sun	57 910 000 km (0.38 AU)
Diameter	4880 km
Mass	3.3×10^{23} kg (0.055 times Earth's mass)
Density	5.43 g/cm³ or 5430 kg/m³
Orbital eccentricity	0.206
Period of revolution (length of year)	88 Earth days or 0.241 Earth years
Rotation period	58.65 Earth days
Orbital velocity	172 400 km/h
Tilt of axis	2°
Day temperature	430°C
Night temperature	−180°C
Number of moons	0
Atmosphere	Practically none (some oxygen, sodium, helium)
Strength of gravity	3.3 N/kg at surface

could see clouds above its surface. We now know that Mercury's atmosphere is far too tenuous to support clouds. The lack of clouds is also due to the fact that Mercury's escape velocity is only 4.3 km/s, so any gas particles would be moving too quickly to be restrained by Mercury's gravity.

PROBING MERCURY

Mercury is the least explored of our solar system's inner planets. To date the planet has been visited by only two spacecraft.

Mariner 10 flew past Venus on 5 February 1974, in order to get a gravity assist to Mercury. It flew by the planet three times between March 1974 and March 1975. Mariner 10 was also the first spacecraft to have an imaging system, and the encounter produced over 10 000 pictures that covered 57 per cent of the planet. Mercury is too close to the Sun to be mapped by the Hubble Space Telescope.

Another NASA spacecraft, called **Messenger**, was launched on a mission to Mercury on 2 August 2004. Messenger stands for 'MErcury Surface, Space ENvironment, GEochemistry and Ranging'. This probe's seven-year journey will include fifteen trips around the Sun, one Earth fly-by, two

Venus fly-bys, and three Mercury fly-bys (January 2008, October 2008, September 2009) before it enters orbit around Mercury in March 2011. The fly-bys will help focus the science mission when the spacecraft enters orbit. The probe is expected to orbit Mercury for one year. With a package of seven scientific instruments, Messenger will determine Mercury's composition, image its surface globally, map its magnetic field, measure the properties of its core, explore the mysterious polar deposits to learn whether ice lurks in permanently shadowed regions, and characterise Mercury's tenuous atmosphere and Earth-like magnetosphere.

Pictures taken by Messenger in January 2008 show the far side of Mercury contains the wrinkles of a shrinking, ageing planet. There are scars from volcanic eruptions, and craters with a series of troughs radiating from them.

Figure 4.2   The Messenger space probe took this photograph of the far side of Mercury in January 2008. The previously unseen features suggest the planet is old and wrinkly. (Photo: NASA)

Table 4.2 Significant probes sent to Mercury

Name of probe	Country of origin	Launched	Comments
Mariner 10	USA	1973	Now in solar orbit
Messenger	USA	2004	First fly-by January 2008; to go into orbit around Mercury in 2011

A joint European–Japanese mission to Mercury is scheduled for launch between 2009 and 2012. The mission, called 'BepiColombo', is made up of two orbiters and a possible lander.

POSITION AND ORBIT

Mercury has the most eccentric orbit of all the planets. At closest approach (perihelion) it has an orbital radius of only 46 million kilometres, but its furthest distance (aphelion) has a radius of 70 million kilometres. At perihelion Mercury travels around the Sun at a very slow rate.

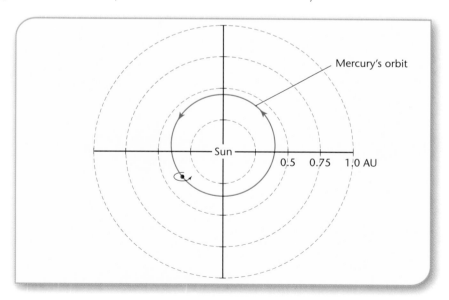

Figure 4.3   Shape of Mercury's orbit (distance circles are in astronomical units, AU).

Observation of Mercury is difficult because of its close proximity to the Sun. The best time to view the planet is when it appears above the horizon twice a year at its greatest distance from the Sun. At these times, Mercury can be seen just before sunrise or just after sunset.

The axis of rotation of Mercury is almost vertical. This means the plane of its equator coincides with the plane of its orbit.

Mercury has the shortest year of any planet, taking only 88 days to orbit the Sun. It has no known satellites (moons). Because it is closer to the Sun than Earth is, Mercury is seen to go through phases just like our Moon. Mercury's size appears to vary according to its phases because of its changing distance from Earth. When Mercury first appears in the evening sky, it is coming around the far side of its orbit toward us, and through a telescope it appears as a thin crescent.

At rare intervals, observers from Earth can see Mercury pass in front of the Sun. Such a passing is called a **transit**. Transits occurred on 7 May 2003 and 8 November 2006. A transit should be viewed by projecting the Sun's image from a telescope onto a white screen. The planet would appear as a black dot slowly moving across the Sun's image. Care should always be taken when viewing the Sun – never look directly at the Sun through a telescope.

DID YOU KNOW?

The high eccentricity of Mercury's orbit would produce strange effects for anyone on the surface. For example, at some longitudes the observer would see the Sun rise and then gradually increase in size as it slowly moved overhead. Once overhead, the Sun would appear to reverse course, then stop again before resuming its path toward the horizon and decreasing in size. At night, stars would appear to move three times faster across the sky. Observers at other places on Mercury's surface would see different but equally strange motions.

DENSITY AND COMPOSITION

Mercury is thought to be one of the densest planets in the solar system. This high density suggests that Mercury is dominated by an iron core, which makes up 70 per cent of the planet by mass. During 2007, a team of astronomers announced that they had evidence suggesting that some of the core is molten. The discovery was made by bouncing radio waves off the planet and analysing the return signals.

A thin layer or mantle surrounds the core, which is mainly silica rock (30%). Little is known about Mercury's crust, but it is thought to extend down less than 100 kilometres. The crust seems to have cooled rapidly once formed, and is solid enough to preserve surface features.

Mercury is about one-third the size of Earth and has a little more than one-third the gravity of Earth. A 75 kg person on Earth has a weight of 735 N, but on Mercury the same person would weigh 247 N.

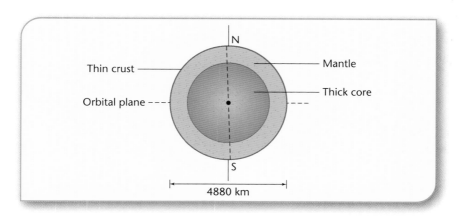

Figure 4.4   Interior of Mercury.

THE SURFACE

The surface of Mercury is heavily cratered and looks very similar to that of our Moon. Most of the craters are impact craters formed from bodies colliding with the surface. The distribution of craters on Mercury, the Moon and Mars are similar, suggesting the same kind of object was responsible for the impacts on each body. One of the largest impact features on Mercury is the Caloris Basin. This basin measures about 1300 km across and has been partially flooded with lava from volcanic activity. It was probably formed by a very large impact early in the history of the solar system. As the basin floor settled under the weight of volcanic material, fractures and ridges formed.

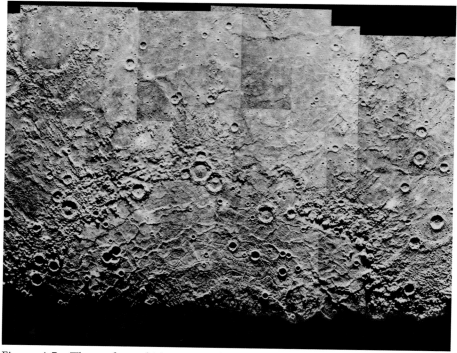

Figure 4.5   The surface of Mercury in the region of the Caloris Basin (bottom) as seen by Mariner 10. Mariner 10 made three fly-bys of Mercury in 1974 and 1975. (Photo: NASA)

In addition to the heavily cratered areas, Mercury also has large regions of smooth plains formed by ancient lava flows. Mariner 10 also revealed some large escarpments over the surface, some up to hundreds of kilometres in length and three kilometres high. Some of these cut through the rings of craters, and others seem to be formed by tectonic forces. The largest scarp or cliff observed to date is Discovery Rupes, which is about 500 kilometres long and 3 kilometres high. Such scarps are thought to have formed as a result of global compression and tectonic activity as Mercury cooled.

Although data collected by Mariner 10 suggests that recent volcanism has occurred on Mercury, many of Mercury's craters are not covered by volcanic flows, indicating that they were formed after volcanic flows ceased. Heavy bombardment of the planet ended about 3.8 million years ago.

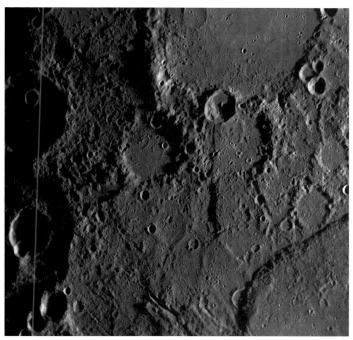

Figure 4.6   Mariner 10 photograph of Santa Maria Rupes. This fault (lower right) runs through many craters, indicating that it formed more recently than the craters themselves. (Photo: NASA)

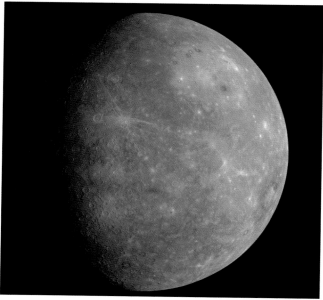

Figure 4.7 Hilly and cratered terrain on Mercury. On the upper right is the Caloris Basin, including its western portions never before seen by spacecraft. (Photo: NASA)

Recent radar analysis has found that a number of craters near each pole have high radar reflectivity, suggesting the presence of ice. The interiors of these polar craters are permanently shaded from the Sun's heat, making the preservation of ice possible.

MERCURY'S ATMOSPHERE

Mercury has a very thin atmosphere. In 1974, Mariner 10 detected traces of oxygen, sodium, helium, potassium and hydrogen vapours. More recently, Earth-based telescopes have detected gaseous sodium, potassium and calcium. The hydrogen and helium may have originated from the solar wind, while the sodium may have come from surface rocks bombarded by the wind or meteorites. Astronomers have observed clouds of sodium vapour occasionally rising from the surface of Mercury.

The atmospheric pressure is only a million-billionth that of Earth's – as low as many vacuums created in Earth laboratories. Mercury has very little atmosphere because its gravity is too weak to retain any significant gas particles, and it is so hot that gases quickly escape into outer space. The quantities of these gases have been poorly determined and vary depending on the position of Mercury in its orbit. The Messenger probe should greatly increase our knowledge of Mercury's atmospheric composition.

The atmospheric gases are much denser on the cold night side of Mercury than on the hot day side.

TEMPERATURE AND SEASONS

Because of its closeness to the Sun, surface temperature variations are extreme on Mercury, more than on any other planet. You could roast during the day at 430°C, and freeze during night at −180°C. The high day temperature would be hot enough to melt the metals zinc and tin. One day on Mercury is equal to 58.65 Earth days, so it takes a long time to warm up from the cool of night and it takes a long time to cool down from the heat of day.

Mercury is not the hottest planet on average. The temperature on Venus is slightly higher than on Mercury but is more stable because of the thick clouds on that planet.

On Earth, the seasons change in a regular way because the rotational axis is inclined (at 23.5° from the perpendicular to its orbital plane). As a result, each hemisphere on Earth receives more direct sunlight during one part of the orbit than the other. Mercury's axis is very nearly perpendicular to its orbital plane (2° from perpendicular), so no seasonal changes occur. Some craters near Mercury's poles never receive any sunlight and are permanently cold.

MAGNETIC FIELD

The magnetic field of a planet is generated by electric currents flowing in a molten metal core as a planet rotates. Because Mercury rotates slowly (59 times more slowly than Earth), it was not expected to have a magnetic field. However, Mariner 10 found that Mercury does have a weak magnetic field, about 1 per cent of that of Earth.

The magnetic poles coincide with Mercury's rotational poles.

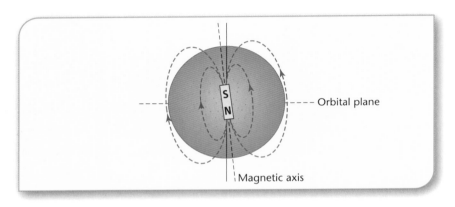

Figure 4.8   Mercury has a weak magnetic field.

In 2011 the Messenger space probe will go into orbit around Mercury and part of its mission is to map Mercury's magnetic field.

WEB NOTES

For fact sheets on any of the planets, including Mercury, check out:
    <http://nssdc.gsfc.nasa.gov/planetary/planetfact.html>
    <http://www.nasm.si.edu/etp/>
    <http://www.space.com/mercury/>
    <http://solarsystem.nasa.gov>

CHAPTER 5

VENUS – A HOT, CLOUDY PLANET

VENUS IS THE second planet from the Sun, orbiting on average at a distance of 108 million kilometres from the Sun. It is the sixth largest planet, with a diameter of 12 104 km. Venus is sometimes regarded as Earth's sister planet since it is similar in size and mass to Earth. Venus orbits the Sun between Mercury and Earth but it is twice as far as Mercury from the Sun. It comes closer to Earth than any other planet in the solar system.

One of the strange things about Venus is that it spins on its axis in the opposite direction to that of the other terrestrial planets – it seems to be upside down, with its north and south poles reversed. The rotation is thought to be due to a massive impact early in the planet's life.

Venus is thought to have formed at the same time as the other planets in the solar system, about 4.5 billion years ago. Because it is close to the Sun, Venus must have been hot and in a molten state before it cooled to become a solid planet. Out of all the planets, Venus still has the highest surface temperature, even though it is not the closest planet to the Sun. This is mainly because of its noxious atmosphere, which pushes down on the surface with a pressure 92 times that experienced on Earth.

Venus has no natural satellites.

Figure 5.1 Venus as seen from the Pioneer spacecraft in 1978. This ultraviolet photograph shows the circulation pattern of the clouds. The clouds move rapidly from east to west across the planet. (Photo: NASA)

EARLY VIEWS ABOUT VENUS

Venus is the brightest planet as seen from Earth. At certain times of the year it can be seen in the evening sky just after sunset; at other times of the year it appears to rise in the east just before sunrise.

Venus was well known to the ancient Greeks and Romans because of its brightness in the night sky, but the Greeks believed Venus to be two different objects: Phosphorus as the morning star, and Hesperos as the evening star.

To the ancient Romans, Venus was the goddess of love and beauty. (Venus is the only planet named after a goddess.) Venus is very bright as seen from Earth because its covering of dense clouds reflects over three-quarters of the sunlight received by the planet. These clouds completely hide the surface of the planet from view.

Table 5.1 Details of Venus

Distance from Sun	108 200 000 km (0.72 AU)
Diameter	12 104 km
Mass	4.87×10^{24} kg (0.82 times Earth's mass)
Density	5.25 g/cm^3 or 5250 kg/m^3
Orbital eccentricity	0.007
Period of revolution (length of year)	224.7 Earth days or 0.615 Earth years
Rotation period	243 Earth days
Orbital velocity	126 108 km/h
Tilt of axis	177.3°
Day temperature	480°C
Night temperature	470°C
Number of moons	0
Atmosphere	carbon dioxide
Strength of gravity	8.1 N/kg at surface

PROBING VENUS

People have long thought that because Venus is close to Earth and similar in size and mass, it would have conditions suitable for life. However, early robotic probes sent to Venus disproved this theory. The probes found Venus to have a hostile-to-life or hell-like environment. The thick atmosphere, high surface temperature and high pressure hampered early exploration by spacecraft, and many probes were unsuccessful. Both the USSR and USA have sent more probes to Venus than to any other planet, mainly because of its closeness to Earth.

During the 1960s the USSR launched a series of Venera spacecraft on missions to Venus. **Venera 1** in 1961 was the first space probe to fly by Venus. Communications with **Veneras 2** and **3** failed just before arrival. The first successful probe to enter the Venusian atmosphere was **Venera 4**, on 18 October 1967. Although this craft was crushed during descent, it sent back useful data on the planet's atmosphere, including its chemical composition, pressure and temperature. In 1969, **Venera 6** returned atmospheric data down to within 26 kilometres of the surface before being crushed by the pressure. The first successful landing of a spacecraft on any planet was achieved by **Venera 7** on 15 December 1970. It used an external cooling device to allow it to send back 23 minutes of data. **Venera 9** included an orbiter and a lander – the lander arrived on the Venusian surface on 22 November 1975 and transmitted the first black-and-white images of the planet's surface. **Venera 13** survived for 2 hours and 7 minutes on the Venusian surface. It took colour images and analysed a soil sample. The first colour panoramic views of the surface were sent back by **Venera 14** in November 1981. This probe also conducted soil analysis using an X-ray fluorescence spectrometer. **Veneras 15** and **16** were the first spacecraft to obtain radar images of the surface from orbit. The images were used to produce a map of the northern hemisphere from the pole to 30° north latitude. During 1985, **Vegas 1** and **2** flew by Venus for a fly-by on their way to comet Halley. Vega 1 dropped off a Venera-style lander and a balloon to investigate the cloud layers. The lander from Vega 1 failed, but Vega 2's lander was able to collect soil samples. Both Vega 1 and Vega 2 are now in solar orbit.

The USA, through NASA, was also active in sending probes to Venus. The first US probe to make a fly-by of Venus was **Mariner 2**, on 14 December 1962. Mariner 2 passed Venus at a distance of 34 800 km, and scanned its surface with infrared and microwave radiometers, showing the surface temperature to be about 425°C. The temperatures were confirmed by **Mariner 5** in 1967 as it passed within 3900 km of the planet. It also studied the magnetic field, and found an atmosphere containing 85–97% carbon dioxide. **Mariner 10** flew past Venus on 5 February 1974 for a gravity assist to the planet Mercury. It recorded circulation in the atmosphere of Venus and showed the temperature of the cloud tops to be –23°C.

In 1978 the USA launched two Pioneer Venus probes to orbit Venus. **Pioneer Venus 1** (also known as Pioneer 12) operated continuously from 1978 until 8 October 1992, when contact was lost and it burnt up in the Venusian atmosphere. The orbiter was the first probe to use radar imaging in mapping the planet's surface. **Pioneer Venus 2** (also known as

Pioneer 13) carried four atmospheric probes that were released on 9 December 1978. The four probes descended by parachute and collected data on the atmospheric layers before burning up in the atmosphere. One of the sub-probes landed intact and sent back data for over an hour.

The **Galileo** spacecraft flew past Venus on its way to Jupiter in February 1990.

The USA also sent the **Magellan** spacecraft into orbit around Venus in August 1990. This probe was launched from the space shuttle Atlantis in May 1989 and took 15 months to reach Venus. Its main mission was to produce a high-resolution map of Venus using synthetic aperture radar, which can see through clouds. The spacecraft mapped 99 per cent of the planet's surface from a polar orbit. In 1994, the craft was directed into the atmosphere where it burned up.

Figure 5.2   The Magellan spacecraft showed Venus is covered with extensive lava flows and lava plains. Numerous images were overlapped to obtain this overall image of Venus. The bright areas are the equatorial highlands known as Aphrodite Terra. (Photo: NASA)

Table 5.2 Significant space probes to Venus

Probe	Country of origin	Launched	Comments
Venera 1	USSR	1961	Now in solar orbit
Mariner 2	USA	1962	Venus fly-by
Zond 1	USSR	1964	Venus fly-by, communication lost en route
Venera 2	USSR	1965	Communication failed before arrival
Venera 3	USSR	1965	Communication failed before atmospheric entry
Venera 4	USSR	1967	Entered atmosphere of Venus, eventually crushed
Mariner 5	USA	1967	Passed close to Venus, returned data, now in solar orbit
Venera 5	USSR	1969	Entered atmosphere, returned data before crashing
Venera 6	USSR	1969	Atmospheric probe
Venera 7	USSR	1970	First successful unmanned landing on another planet
Venera 8	USSR	1972	Lander on surface, returned data for 50 minutes
Mariner 10	USA	1973	Venus fly-by in 1974, first craft to have an imaging system
Venera 9	USSR	1975	Transmitted first black-and-white photographs of the surface
Venera 10	USSR	1975	Transmitted black-and-white images of the surface
Pioneer Venus 1	USA	1978	Mapped surface using radar, operated until October 1992
Pioneer Venus 2	USA	1978	Released four atmospheric probes
Venera 11	USSR	1978	Returned data for 95 minutes before failing
Venera 12	USSR	1978	Landed; returned data for 110 minutes
Venera 13	USSR	1981	Returned black-and-white and colour images of the surface, performed soil analysis
Venera 14	USSR	1981	Returned black-and-white and colour images of the surface, performed soil analysis
Venera 15	USSR	1983	Produced a map of northern hemisphere
Venera 16	USSR	1983	Further mapping of northern hemisphere
Vega 1	USA	1984	Venus fly-by, lander's soil experiment failed
Vega 2	USA	1984	Venus fly-by, lander sampled soil
Galileo	USA	1989	Venus fly-by en route to Jupiter
Magellan	USA	1989	Mapped 99 per cent of Venus using radar during 1990–94
Venus Express	ESA	2005	Reached Venus in 2006 – polar orbiter

The **Venus Express** probe was launched by the European Space Agency on 9 November 2005. On 11 April 2006 the probe went into polar orbit around Venus. At closest approach, the probe was about 250 km above the north pole and 66 000 km above the south pole. The main objectives of the mission include exploring the global circulation of the Venusian atmosphere, chemistry of the atmosphere, surface volcanism, and atmospheric loss. Thermal imaging done by the probe showed a thick layer of clouds, located at about 60 km altitude, that traps heat radiating from the surface.

POSITION AND ORBIT

Venus orbits the Sun in a nearly circular orbit, as shown by its small orbital eccentricity. Its mean orbital radius is just over 108 million kilometres, and it passes within 40 million kilometres of Earth, closer than any other planet.

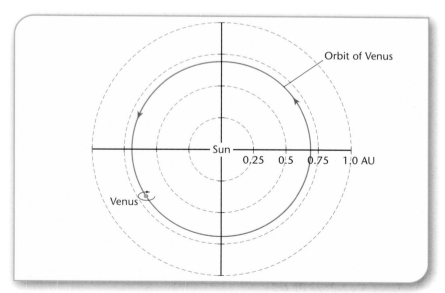

Figure 5.3   Orbital path of Venus (distance circles are in astronomical units, AU).

Observation of Venus is easy because of its close proximity to Earth, and it is the brightest object in the sky apart from the Sun and Moon. Venus can be seen either in the eastern sky before sunrise or in the western sky after sunset. It is so bright it can often be seen during daylight and is often mistaken for a UFO. Cloud cover on Venus is 100 per cent, compared to an average of 40 per cent on Earth; so optical telescopes on Earth cannot be used to see the surface of Venus.

Venus orbits the Sun in about 225 Earth days. Because it is closer to the Sun than Earth, Venus is seen to go through phases just like our Moon. Venus's size appears to vary according to its phases because of its changing distance from Earth. Venus is in its full phases when furthest from Earth, and when close to Earth it is seen as a thin crescent phase. It is possible to view the phases through binoculars or a small telescope.

As Venus and Earth travel around the Sun, Venus can be seen near the opposite side of the Sun about every 584 days. When Venus is moving towards Earth, the planet can be seen in the early evening sky (west). When Venus is moving away from Earth, the planet can be seen in the early morning sky (east).

Figure 5.4 The 2004 transit of Venus across the southern part of the Sun. (Photo: J. Wilkinson)

At rare intervals, observers from Earth can see Venus transit, or pass in front of, the Sun. The last transit occurred on 8 June 2004, and the next is due on 6 June 2012. A transit should be viewed by projecting the Sun's image onto a white screen from a telescope. The planet can be seen as a black dot slowly moving across the Sun's image (Figure 5.4). Care should always be taken when viewing the Sun – never look directly at the Sun through a telescope.

DENSITY AND COMPOSITION

Venus and Earth have almost the same mass, diameter and average density.

The strength of gravity on Venus is slightly less than that on Earth. A 75 kg person on Earth weighs 735 N, but on Venus they would only weigh 607 N. Venus is the third densest planet in the solar system. This high density is because Venus has a large rocky core made of mostly nickel and iron. The core is about 3340 km in radius and is surrounded by a molten silicate mantle about 2680 km thick. There is also a thin outer layer or crust about 50 km thick that is similar to the crust on Earth.

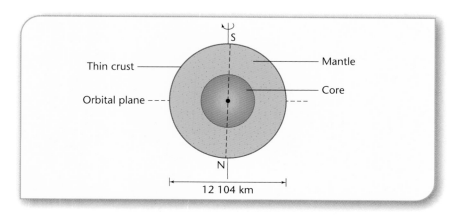

Figure 5.5 The interior structure of Venus.

THE SURFACE

Although the surface of Venus cannot be seen from Earth because of the thick clouds surrounding the planet, some features have been detected by Earth-based radar.

In the 1960s, both the USSR and USA began sending space probes to Venus. Reaching the surface proved to be more difficult than anyone expected because the atmospheric pressure was so great that it crushed many early craft.

Instruments on Venera 7 found the surface temperature was 475°C and the surface pressure was 90 atmospheres (about the same as the pressure at a depth of 1 km in Earth's oceans). The black-and-white photographs returned from the Venera 9 lander showed a rocky terrain with basaltic stones several centimetres across and soil scattered between them. The temperature at the landing site was 460°C and wind speed was only 2.5 km/h. The terrain around the lander from **Venera 10** was more eroded than at the Venera 9 landing site.

Soil analysis by **Venera 14** showed the surface rock type to be basaltic, similar to that found at mid-ocean ridges on Earth.

Veneras 15 and **16** produced a map of the northern hemisphere from the pole to 30°N and found several hot spots that possibly were caused by volcanic activity. Most of the surface of Venus consists of gently rolling plains with no abrupt changes in topography. There are also several broad depressions called lowlands, and some large highland areas including two named Aphrodite Terra and Ishtar Terra. The highland areas can be compared to Earth's continents, and the lowland areas to its ocean basins. Several canyons and a few rift valleys were also mapped.

There are no small craters on Venus, probably because small meteoroids burn up in Venus's dense atmosphere before reaching the surface. There are some large craters and these appear to come in bunches, suggesting that large meteoroids break up into pieces just before hitting the ground.

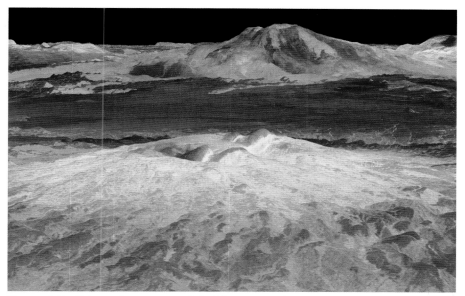

Figure 5.6   False-colour image of the Venusian volcano Sapas Mons as produced by Magellan. Sapas Mons is approximately 400 km across and 1.5 km high. (Photo: NASA)

The best pictures of the Venusian surface came from the orbiting US spacecraft **Magellan**, which produced detailed maps of Venus's surface using radar. Images from Magellan confirmed that that much of the surface is covered by flat rolling plains with several highland regions. The highest mountain on Venus is Maxwell Montes, situated near the centre of Ishtar Terra and rising to around 11 km above the mean surface level. Maxwell Montes is about 870 km long.

Recent data suggests that Venus is still active volcanically, but only in a few hot spots. There is evidence of lava flows, volcanic domes, collapsed volcanic craters and volcanic plains. There are also several large shield type volcanoes (similar to those at Hawaii).

Two large and possibly active shield volcanoes are Rhea Mons and Theia Mons, which tower 4 km high.

DID YOU KNOW?

Venusian volcanoes tend to be broad but not as high as those on Earth. The highest volcano on Venus is Maat Mons, which rises 8.5 km above the surrounding plains. By comparison, the volcanoes on Hawaii rise 10 km above the sea floor.

Radar images from Magellan suggest there have been recent lava flows on Venus. Such volcanic activity provides evidence about the interior processes of a planet. Venus appears to lose heat from its interior via hot-spot volcanism rather than via convection as in the case of Mercury. Hot spots produce shield volcanoes and flood volcanism, and these features are common on the present-day Venusian landscape. Magellan also revealed surface features called arachnoids, which look like craters with spider legs radiating from them. These features are thought to form when molten magma pushes up from the interior with such force that the surrounding crust gets cracked.

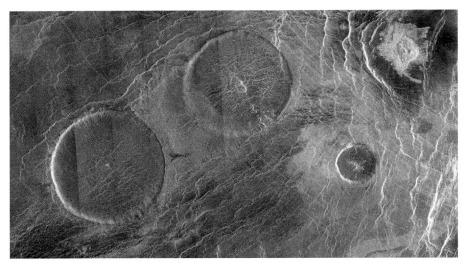

Figure 5.7 The Magellan probe found four distinctive 'pancake domes' on the eastern edge of the Alpha Regio highland plateau. These volcanic features are about 25 km in diameter and 750 m high. (Photo: NASA)

Tectonic movement that has resulted in crustal shortening, stacking of crustal blocks and wrinkle ridges on the lowlands and rolling plains has also shaped the surface of Venus. Also seen on the lowlands and plains are fractures formed when the crust was stretched or pulled apart. Diana Chasma is the deepest fracture on Venus, with a depth around 2 km below the surface and a width of nearly 300 km.

The oldest terrains on Venus are about 800 million years old. Lava flows at that time probably wiped out earlier surface features and larger craters from early in Venus's history.

Venus probably once had large amounts of water like Earth but the high temperature boiled all this away, so Venus is now very dry.

THE ATMOSPHERE

Many of the probes sent to Venus have provided information about its atmosphere. The clouds of Venus conceal a hostile atmosphere that reaches a height of about 250 km. Most of the atmosphere is concentrated within 30 km of the surface.

The first probe to be placed directly into the atmosphere and to return data was Venera 4 in 1967. It showed that the atmosphere was 90–95 per cent carbon dioxide. Mariner 5 arrived at Venus one day after Venera 4, and passed within 3900 km of the planet's surface – it also found an atmosphere dominted by carbon dioxide. Veneras 5 and 6 reported an atmosphere of 93–97 per cent carbon dioxide, 2–5 per cent nitrogen, and less than 4 per cent oxygen. These two probes returned data to within 26 km and 11 km of the surface respectively before being crushed by the high atmospheric pressure.

From 1978 to 1988, the amount of sulfur dioxide in the atmosphere decreased by 10 per cent. The reason for this decrease may have been a decrease in volcanic activity during this period.

In 1978, the Pioneer Venus 2 probe detected a fine haze in the atmosphere at a height of 70–90 km. Between 10–50 km there was some atmospheric convection, and below 30 km the atmosphere was clear.

Unlike the clouds on Earth, which appear white from above, the cloud tops on Venus appear yellowish or yellow-orange. These colours are thought to be caused by sulfur and sulfur compounds in the atmosphere. Evidence suggests these compounds have originated from volcanic activity.

TEMPERATURE AND SEASONS

Surface temperatures on Venus can rise to 480°C, hot enough to melt lead, zinc and tin. The high temperature and pressure were responsible for the failure of many early space probes. Temperature and pressure in the atmosphere decrease with increasing altitude.

The dense atmosphere on Venus allows heat from the Sun to warm the surface but it also traps heat radiated from the surface of Venus. This results in a higher surface temperature than on Mercury (which is closer to the Sun). The trapping of heat by the atmosphere produces a greenhouse effect because the carbon dioxide acts like glass in a greenhouse. Earth has a greenhouse effect in its atmosphere, but Venus is an extreme case. The thick atmosphere of Venus also keeps the night side of Venus at nearly the same temperature as the side facing the Sun (unlike the night side of Mercury, where the temperature drops dramatically). Temperatures at the poles of Venus are as high as those at the equator.

As it orbits the Sun, Venus rotates very slowly on its axis, more slowly than any other planet. It takes 243 Earth days for just one spin, which means that a Venusian day is longer than a Venusian year.

Venus's rotation axis is tilted more than 177°, compared to Earth's 23.5° tilt. This means that Venus's axis is within 3° of being perpendicular to the plane of its orbit around the Sun. Because of this, the planet has no seasons.

Neither of the planet's hemispheres or poles point towards the Sun during any part of its orbit.

Magnetic field

The volcanic activity on Venus suggests it has a molten interior that should be able to generate electric currents. The core of Venus is also thought to be mainly iron, like Earth's core. Both these facts suggest Venus should have a magnetic field. However, none of the spacecraft sent there has detected one. The lack of a magnetic field may be a consequence of the slow rotation of Venus on its axis.

Web notes

For fact sheets on any of the planets, including Venus, check out:
    <http://nssdc.gsfc.nasa.gov/planetary/planetfact.html>
    <http://www.space.com/venus/>
    <http://www.nasm.si.edu/etp/>

CHAPTER 6

EARTH –
A WATERY, LIVING WORLD

EARTH IS THE third planet from the Sun, with an average orbital radius of 150 million kilometres. This distance is also known as one astronomical unit (1 AU).

Earth is the fifth largest planet, with a diameter of 12 756 km. This makes Earth slightly larger than Venus. Earth orbits the Sun between Venus and Mars, and is 1.4 times further from the Sun than is Venus.

Earth is thought to have formed at the same time as the other planets in the solar system, about 4.5 billion years ago. Scientists know the age of the Earth because the oldest rocks ever discovered are 4.3 million years old (determined from radioactive dating). Earth is the largest of the four rocky or terrestrial planets (Mercury, Venus, Earth and Mars). Like Mercury and Venus, Earth formed from a hot and molten state before it cooled to become a solid planet. Even though it has cooled since formation, Earth is currently the most geologically active planet, and its interior is still hot. The main feature that separates Earth from the other planets in the solar system is that it is the only planet to contain water in the liquid state.

Figure 6.1 The Earth as seen from a satellite about 36 000 km out in space. Water covers about 70 per cent of the Earth. Land covers only about 30 per cent. (Photo: NASA)

Table 6.1 Details of Earth

Distance from Sun	149 600 000 km (1.0 AU)
Diameter	12 756 km
Mass	5.97×10^{24} kg
Density	5.52 g/cm^3 or 5520 kg/m^3
Orbital eccentricity	0.017
Period of revolution (length of year)	365.3 Earth days or 1.00 year
Rotation period	1.0 Earth days (24 hours)
Orbital velocity	107 244 km/h
Tilt of axis	23.5°
Day temperature	15°C
Night temperature	10°C
Number of moons	1
Atmosphere	Nitrogen, oxygen
Strength of gravity	9.8 N/kg at surface

Early views about Earth

Earth is the only planet whose English name has not been derived from Greek or Roman mythology. The name comes from Old English and Germanic. There are, of course, names for Earth in every language.

Ancient understandings of Earth varied often according to religious views. For a long time people thought the Earth was flat because it seemed that way. It was not until the time of Copernicus in the sixteenth century that it was understood that the Earth was just another planet.

Probing Earth

Human exploration of the solar system began with Earth. The first spacecraft were small unmanned craft launched into Earth's atmosphere. Improvements in space technology led to manned craft orbiting Earth.

Much of Earth has been studied without the aid of spacecraft. However, data from spacecraft confirmed much of what was known and enabled more accurate maps to be made. Pictures taken by spacecraft from space around Earth have helped scientists to predict weather and track hurricanes, for example.

Early in the space race, the USSR was active with its Sputnik, Vostok, Voskhod and Soyuz spacecraft. Early US spacecraft placed in Earth orbit included Explorer and those of the Mercury and Gemini programs. These early missions were the pioneers of future exploration of the Moon and other planets. For example, the 10 manned Gemini missions between 1964 and 1966 involved rendezvous between spacecraft in orbit, space walks, and dockings with unmanned target vehicles. The American Apollo program featured many spacecraft missions that eventually resulted in humans visiting the Moon (Earth's only natural satellite).

Today there are many artificial satellites in orbit around Earth. These are being used for many purposes, including communications, defence, science, navigation (including GPS) and weather monitoring. Orbiting

Table 6.2 Significant space probes used to monitor Earth

Probe	Country of origin	Launched	Comments
Explorer 1	USA	1958	Cosmic rays monitor, discovered Van Allen radiation belts
Vanguard II	USA	1959	First satellite to send weather data back to Earth
Tiros 1	USA	1960	Took first weather pictures
Explorer XVII	USA	1963	First satellite to study atmosphere
Landsat 1	USA	1972	Surveyed Earth's natural resources
SMS1	USA	1974	First full-time weather satellite in synchronous orbit with Earth
Lageos	USA	1976	First satellite to make high-precision geographic measurements
Goes 3	USA	1978	Day and night photographs of weather
Seasat	USA	1978	Used radar to examine Earth and seas
Nimbus	USA	1978	Monitors oceans and atmosphere
Sage	USA	1979	Studied composition of atmosphere
Solar Mesosphere Explorer	USA	1981	Monitored the ozone layer
Shuttle Imaging Radar	USA	1981, 1984, 1994, 2000	Mapped Earth's topography
Topex/Poseidon	USA, France	1992	Measures sea levels every 10 days
NASA Scatterometer	USA	1996	Monitors ocean currents and winds
Quick Scatterometer	USA	1999	Monitors ocean currents and winds
Active Cavity Irradiance Monitor Satellite	USA	1999	Monitors the Sun's energy
Terra	USA	1999	Photographs Earth in 14 colour bands
Image	USA	2000	Weather satellite; monitors space storms and magnetosphere
Jason 1	USA	2001	Monitors ocean circulation and atmospheric conditions
Aqua	USA	2002	Accurately monitors temperatures, humidity, clouds
Grace	USA, Germany	2002	Measures gravity, surface currents, ocean heat
Aura	USA	2004	Monitors troposphere

Figure 6.2   One of the uses of the Space Shuttle is to launch satellites into orbit around Earth. The satellites are carried into space by the shuttle and released through the cargo doors via a robotic arm. (Photo: NASA)

observatories, such as Skylab, Mir, and the International Space Station, are the largest and most complicated of all scientific satellites. Many types of instruments can be built into each one.

POSITION AND ORBIT

Earth orbits the Sun in a slightly elliptical orbit. Its mean distance from the Sun is just over 149 million kilometres. At closest approach (perihelion) it is 147 million kilometres from the Sun, and its greatest distance (aphelion) is 152 million kilometres.

Earth takes just over 365 days, 6 hours and 9 minutes to orbit the Sun once; this length of time is called a **sidereal** year. Its axis is inclined at 23.5° off the perpendicular to the plane of its orbit. A single rotation on this axis takes 23 hours 56 minutes and 4 seconds; this length of time is called a

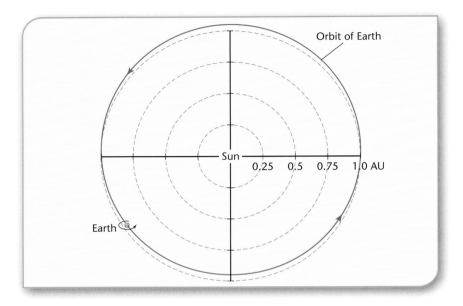

Figure 6.3 Orbit of Earth around the Sun (distance circles are in astronomical units, AU).

sidereal day. Earth rotates from west to east on its axis and this makes the Sun appear to move across the sky from east to west each day. The tilt of Earth's axis causes the seasons.

The interaction of Earth and the Moon slows Earth's rotation by about 2 milliseconds per century. Current research indicates that about 900 million years ago there were 481 eighteen-hour days in a year.

DENSITY AND COMPOSITION

Earth has a slightly larger mass, diameter and average density than Venus. Because of this, the strength of gravity on Earth is slightly more than that of Venus. A 75 kg person on Earth weighs 735 N. Earth is also the densest planet in the solar system because it has a large nickel–iron core.

Scientists know a lot about the interior of Earth from their study of earthquake (seismic) waves and volcanoes. Earthquakes release enormous amounts of energy that travel through Earth and along its surface as waves or vibrations.

The interior of Earth is divided into three main layers with distinct chemical and seismic properties: the core, mantle and crust. The core has an inner and outer layer. The inner core makes up only 1.7 per cent of Earth's total mass and is solid iron and nickel. It is extremely hot (about 5000°C) but remains a solid because of the pressure from surrounding layers. At the centre the pressure is about four million times greater than at the surface. Surrounding the inner core is a liquid outer core that contributes about 30 per cent of Earth's mass. The outer core has a temperature of about 4100°C. Convection currents in the liquid outer core are thought to produce electrical currents that generate Earth's magnetic field.

The outer core is surrounded by the mantle, which contains 67 per cent of Earth's mass. The mantle is 2820 km thick and contains iron and magnesium silicate minerals and oxygen that are kept solid by high pressures. Nearer the surface of the mantle there are some molten regions that are liquid enough to flow. Sometimes this molten material rises to the surface and forms volcanoes and lava flows.

Surrounding the mantle is a thin outer layer or crust about 40–70 km thick that accounts for about 0.4 per cent of Earth's mass. It is made of granitic and basaltic rocks, which contain silicon dioxide (quartz) and other silicates like feldspar. The surface of the crust is covered by oceans (70%) and continental land (30%).

Table 6.3 Chemical composition of the Earth

Element	Proportion (%)
Iron	34.6
Oxygen	29.5
Silicon	15.2
Magnesium	12.7
Nickel	2.4
Sulfur	1.9
Titanium	0.05

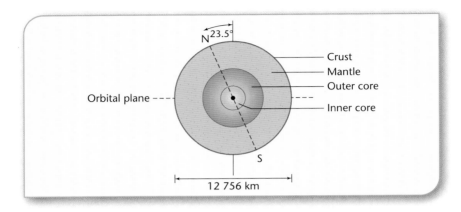

Figure 6.4 The interior structure of Earth.

The crust varies in thickness. It is thicker under the continents and thinner under the oceans.

Unlike the other terrestrial planets, Earth's crust is divided into several separate solid plates, and these move around on top of the hot mantle material. The continents are attached to the plates. Some plates are moving towards each other; when they collide, material builds up to form fold mountains (such as the Himalayas). When plates are moving apart, molten material from the mantle often gets drawn up, creating new crust; this occurs along the Mid-Atlantic Ridge between South America and Africa. Other plates are sliding past each other in a slow but jerky motion; the San Andreas fault in California is one example. Sometimes an oceanic plate slides underneath a continental plate, causing a deep trench (subduction zone) and mountain ranges; the Andes Mountains in South America were formed this way. Where plates interact, earthquake and volcanic activity occurs along the boundaries. Some volcanoes and earthquakes do not occur along plate boundaries, but in places called 'hot spots' – this accounts for the occasional earthquake in Australia, which rides in the middle of a plate.

The slow movement of plates (a few centimetres a year) is driven by convection currents and **tectonic forces** in the mantle.

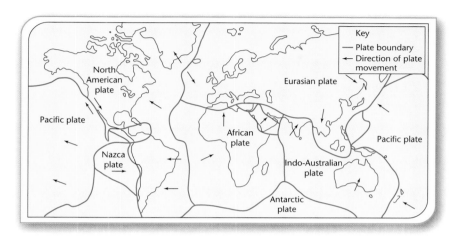

Figure 6.5   Map showing the major plate boundaries of Earth.

Figure 6.6   Aerial photograph of the San Andreas fault about 450 kilometres south of San Francisco. This fault is caused by two plates sliding past each other. Earthquakes occur along this fault line each year. Fortunately, most of the earthquakes are minor and cause little damage, but a large one may not be far away. (Photo: J. Wilkinson)

The surface

The Earth's crust is a thin layer containing rock material.

The surface of the crust consists of mountain ranges, valleys, flat plains, deserts, and vegetated areas of varying density. From space, it can be seen that oceans of water cover much of the surface. The surface has been shaped over millions of years by tectonic forces and volcanic action and erosion by wind, water and glaciers.

The four terrestrial planets – Mercury, Venus, Earth and Mars – contain evidence of volcanic action having occurred in their past. In the case of Earth, some volcanoes are still active and these often erupt, releasing molten rock material into the air or surrounding areas. There are about 500 volcanoes on Earth, most of which are found along plate boundaries.

A key feature of the Earth is the presence of liquid water on its surface. This water is found in the oceans, lakes, and rivers. Water also occurs in the polar ice-caps, and as vapour in the air. Water allows living things to survive on Earth.

Figure 6.7   Volcanic eruption on Earth. (Photo: J. Wilkinson)

DID YOU KNOW?

The highest land feature on Earth is Mount Everest, at 8848 metres above sea level. The lowest land feature is the shore of the Dead Sea, at 399 metres below sea level. The deepest part of the ocean is an area in the Mariana Trench in the Pacific Ocean south-west of Guam, 11 030 metres below the surface. Average ocean depth is 3795 metres.

The largest impact structure discovered on Earth is the Chicxulub Crater. It is hidden under sediments on the coast of the Yucatan Peninsula, Mexico. The crater is a circular structure about 180 km wide and was discovered when instruments detected variations in the Earth's gravitational and magnetic field. The meteorite that made the structure is estimated to have been about 10 km across.

THE ATMOSPHERE

Earth's atmosphere contains 77 per cent nitrogen, 21 per cent oxygen, with traces of argon, carbon dioxide and water.

In recent years the amount of carbon dioxide in the atmosphere has been increasing, due mainly to the burning of fossil fuels. At the same time people are cutting down forest trees that use carbon dioxide. The increase in carbon dioxide levels has resulted in higher average temperatures via the greenhouse effect. Carbon dioxide gas allows sunlight to reach the surface of Earth but prevents heat from escaping, and in this way the atmosphere warms up. However, without the greenhouse effect surface temperatures would be much lower than they currently are.

Oxygen began to accumulate in the atmosphere when primitive life-forms (the first bacteria and plants) began to photosynthesise. Increases in oxygen levels allowed more forms of life to evolve.

The atmosphere of Earth has a number of layers (see Figure 6.8). Each layer corresponds to changes in pressure and temperature.

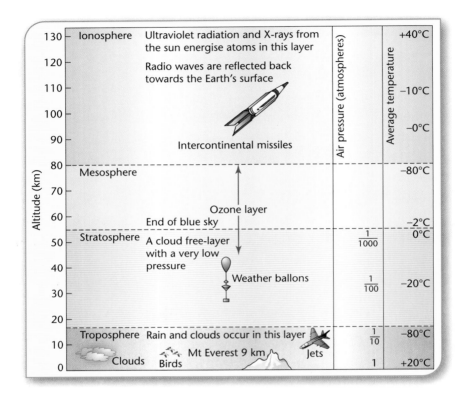

Figure 6.8   Layers in the Earth's atmosphere.

Only a small amount of the heat given off by the Sun enters the atmosphere; most is lost in space. About 34 per cent of the sunlight that enters the atmosphere is reflected back into space by clouds. Only 19 per cent of the sunlight that enters the atmosphere heats it directly. The remaining 47 per cent heats the ground and seas. Re-radiated heat from the ground and seas causes most of the warming of the atmosphere.

Temperature and seasons

Earth is heated by the Sun and has an average surface temperature of 15°C. The highest recorded temperature on the surface is 58°C at Al

Aziziyah in Libya; the lowest temperature recorded is –89.6°C at Vostok Station in Antarctica.

Varying temperatures and pressures in the atmosphere create strong winds.

The Earth has seasons because its axis is tilted at 23.5° off the perpendicular to the plane of its orbit (Figure 6.9).

The four climatic seasons are called summer (hot conditions), autumn (mild conditions), winter (cold conditions), and spring (mild conditions). The seasons in the northern hemisphere are opposite to those in the southern hemisphere. Today the Earth has a fairly stable climate within a narrow range of temperatures. In the past, however, there may have been rapid and dramatic climate change. Such changes may have resulted from changes in the position of the Earth in its orbit around the Sun, increased volcanic activity, changed atmospheric composition, or large meteorite impacts.

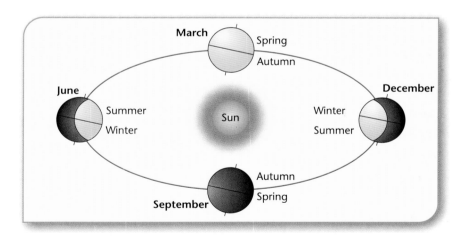

Figure 6.9   The seasons on Earth.

MAGNETIC FIELD

The Earth has a magnetic field produced by electric currents in its iron and nickel core. The magnetic field forms a set of complete loops similar in shape to those around a bar magnet.

The magnetic poles of the Earth are close to the rotational pole (about 11° apart); other planets have much larger angles between their magnetic and rotational poles. The Earth's magnetic field (magnetosphere) normally protects the planet from bombardment by energetic particles from space. Most of these particles originate from the Sun and are carried towards Earth by the solar wind. Near the Earth the solar wind has a speed of about 400 km/s. Charged particles in the solar wind get trapped in the magnetic field and circulate in a pair of doughnut-shaped rings called the Van Allen radiation belts. These belts were discovered in 1958 during the flight of America's first successful Earth-orbiting satellite and are named after physicist James Van Allen, who insisted the satellite carry a Geiger counter to detect charged particles. The outer belt stretches from 19 000 km in altitude to 41 000 km; the inner belt lies between 13 000 km and 7600 km in altitude.

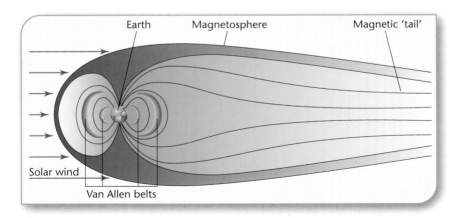

Figure 6.10   The Earth's magnetosphere, showing the Van Allen belts.

During times of solar flares and increased sunspot activity, the Van Allen radiation belts become overloaded with charged particles. Some of the particles travel down into the Earth's upper atmosphere, where they interact with gases like nitrogen and oxygen and make them fluoresce (give off light). The result is a shimmering display called the northern lights (aurora borealis) or southern lights (aurora australis), depending on the hemisphere involved. Such events sometimes also interfere with radio transmissions, communication satellites and electrical power transmission.

THE MOON

Earth has only one natural satellite, the Moon. The Moon orbits Earth in an elliptical orbit at about 36 800 km/h. Its orbital radius varies from 356 000 km to 407 000 km. The Moon is about a quarter the size of Earth, and it has just more than 1 per cent of the mass of Earth. The density of the Moon is only 3.34 g/cm^3 compared to Earth's 5.52 g/cm^3.

Figure 6.11   The Moon as seen from the Earth. (Photo: J. Wilkinson)

The Moon is the brightest object in the sky after the Sun. Unlike the Sun, which emits its own light and heat, the Moon only reflects sunlight. The Moon is also one of the most widely studied objects in the solar system. It has been studied with the naked eye, telescopes and spacecraft. To date, the Moon is the only body in the solar system to have been visited by humans; this occurred for the first time in 1969.

Some scientists believe that the Moon was a body formed at the same time as other planets in the solar system and gravitationally captured by the Earth. Others think it is a fragment torn out of Earth's mantle. Yet another theory is that the Earth–Moon pair could be a double planet.

Early views about the Moon

The presence of the Moon in our sky has captured human interest throughout history. The Moon is so large and close to Earth that some of its surface features are readily visible to the naked eye. For centuries people have thought that some of the features on the Moon looked like a human face looking down on them, and often talked about the 'man in the Moon'. People also thought that the Moon had a fairly smooth surface, until Galileo's telescope showed the surface was covered with many craters and mountains as well as plains.

Table 6.4 Details of the Moon

Distance from Earth	384 400 km
Diameter	3476 km
Mass	7.35×10^{22} kg (0.012 times Earth's mass)
Density	3.34 g/cm^3 or 3340 kg/m^3
Orbital eccentricity	0.055
Period of revolution	27.3 Earth days (relative to stars)
	29.5 Earth days (seen from Earth)
Rotation period	27.3 Earth days
Orbital velocity	36 800 km/h
Tilt of axis	6.7°
Day temperature	130°C
Night temperature	−184°C
Atmosphere	None
Strength of gravity	1.7 N/kg at surface

Probing the Moon

More spacecraft have been sent to the Moon than any other body in the solar system, simply because it is so close to Earth. Early probes to pass by the Moon included the USSR's Luna probes and the USA's Ranger and Surveyor probes. (See Chapter 1 for details.)

The most significant missions to the Moon were those of the USA's Apollo program. The first manned lunar fly-around involved 10 orbits of the Moon by **Apollo 8**, between 21 and 27 December 1968. **Apollo 11** was the first manned lunar landing, on 20 July 1969. The landing site was Mare Tranquillitatis. Neil Armstrong was the first astronaut to walk on the Moon, followed by Edwin Aldrin. The third crew member, Michael Collins, stayed in the command module orbiting the Moon.

Figure 6.12 View of the Moon's surface, the Apollo 11 lander, and distant Earth. (Photo: NASA)

Apollo missions 12 and 14 through to 17 involved landing manned craft on the moon's surface (see Chapter 1 for details). Of these missions, Apollo 15 was the first to make use of a lunar rover vehicle. The last Apollo mission (Apollo 17) occurred in December 1972.

Figure 6.13   Apollo astronaut on the Moon's surface. (Photo: NASA)

Exploration of the Moon continued with the Galileo and Clementine spacecraft. In October 1989, **Galileo** began a six-year mission to Jupiter, but on its way it passed the Moon twice. Data was returned to Earth on the composition of the lunar surface. **Clementine** went into orbit around the Moon in February 1994. Using laser-ranging techniques and high-resolution cameras, Clementine mapped the Moon's topography in greater detail than had been done previously. In 1999, **Prospector** also mapped the lunar surface and detected ice buried beneath the ground in deep polar craters. More recently Japan and China have sent space probes to the Moon.

Table 6.5 Significant space probes sent to the Moon

Probe	Country of origin	Launched	Comments
Pioneer 0, 1, 3	USA	1958	Exploded or failed to reach escape velocity
Luna 1	USSR	1959	First fly-by of Moon
Pioneer 4	USA	1959	Distant fly-by of Moon
Luna 2	USSR	1959	First spacecraft to impact the surface of the Moon
Luna 3	USSR	1959	Took first images of far side of Moon
Ranger 4	USA	1962	First US spacecraft to impact the surface of the Moon
Ranger 5	USA	1962	Fly-by of Moon after landing not possible
Luna 4	USSR	1963	Fly-by after missing Moon
Ranger 6	USA	1964	Probe hit lunar surface, cameras failed
Ranger 7	USA	1964	Sent back pictures before hitting surface
Ranger 8	USA	1965	Returned high resolution pictures before hitting surface
Ranger 9	USA	1965	Returned pictures of its impact on surface
Luna 5	USSR	1965	Lander failed and probe hit surface
Luna 6	USSR	1965	Missed Moon, now in solar orbit
Zond 3	USSR	1965	Returned pictures of far side of Moon, now in solar orbit
Luna 7	USSR	1965	Failed and impacted on surface
Luna 8	USSR	1965	Failed and hit surface
Luna 9	USSR	1966	Landed and returned first pictures from the surface
Luna 10	USSR	1966	Still in lunar orbit
Surveyor 1	USA	1966	First US soft landing on surface
Lunar Orbiter 1	USA	1966	Photographed far side, then hit surface
Luna 11	USSR	1966	In lunar orbit
Surveyor 2	USA	1966	Failed and hit surface
Luna 12	USSR	1966	In lunar orbit
Lunar Orbiter 2	USA	1966	Photographed far side for potential landing sites
Lunar 13	USSR	1966	Landed on lunar surface, used mechanical soil probe
Lunar Orbiter 3	USA	1967	Photographed far side for landing sites, hit surface
Surveyor 3	USA	1967	Landed on lunar surface, examined future Apollo 12 landing site
Lunar Orbiter 4	USA	1967	Orbited over poles of Moon, hit surface
Surveyor 4	USA	1967	Lander failed and hit surface
Explorer 35	USA	1967	Orbiter collected field and particle data
Lunar Orbiter 5	USA	1967	Orbited over poles of Moon, took pictures, hit surface
Surveyor 5	USA	1967	Landed on surface of Moon, analysed soil for magnetism
Surveyor 6	USA	1967	Landed on and took off from lunar surface
Surveyor 7	USA	1968	Landed on surface of Moon, analysed soil near Tycho crater

Probe	Country of origin	Launched	Comments
Luna 14	USSR	1968	In lunar–solar orbit
Zond 5	USSR	1968	Flew around Moon and returned to Earth
Zond 6	USSR	1968	Flew around Moon and returned to Earth
Apollo 7	USA	1968	Earth orbit
Apollo 8	USA	1968	First manned lunar fly-by, 10 orbits of Moon, then returned to Earth
Apollo 10	USA	1969	Manned fly-around, lunar lander descended to near surface
Luna 15	USSR	1969	Unsuccessful sample return attempt, crashed during landing
Apollo 11	USA	1969	First landing and walk by humans on Moon
Zond 7	USSR	1969	Lunar fly-around and return to Earth
Apollo 12	USA	1969	Manned landing, rocks collected, experiments left on Moon
Apollo 13	USA	1970	Intended to be manned landing but explosion en route forced mission abort
Luna 16	USSR	1970	Landed, rock samples returned
Luna 17	USSR	1970	Remote-controlled rover operated for 11 months on surface
Apollo 14	USA	1971	Manned landing, rocks collected using a trolley for transport
Apollo 15	USA	1971	Manned landing, lunar rover vehicle used
Luna 18	USSR	1972	Unsuccessful sample return mission, crashed during landing
Luna 19	USSR	1972	Lunar orbiter
Luna 20	USSR	1972	Robotic landing, returned samples to Earth
Apollo 16	USA	1972	Manned landing, used rover to collect rocks, left instruments
Apollo 17	USA	1972	Manned landing, 75-hour stay, rover driven 30.5 km
Luna 21	USSR	1973	Lunar landing, Lunokhod 2 rover operated for 4 months
Luna 22	USSR	1974	Lunar orbiter
Luna 23	USSR	1974	Crashed on lunar surface
Luna 24	USSR	1976	Landed, samples returned to Earth
Galileo	USA, ESA	1989	Passed Moon in 1990 and 1992 on way to Jupiter
Muses A	Japan	1990	Two orbiters, failed to send back data
Hiten	Japan	1993	Crashed into Fumerius region
Clementine	USA	1994	Lunar orbit, made topographic map of Moon over 70 days
Lunar Prospector	USA	1998	Polar orbit, measured magnetic and gravitational fields

Probe	Country of origin	Launched	Comments
Smart 1	USA	2002	Lunar orbiter designed to test spacecraft technology
Lunar A	Japan	2003	Three probes to surface to study seismology and thermal features
Selene (Kaguya)	Japan	2007	Lunar orbiter and two satellites
Chang'e 1	China	2007	Polar orbit

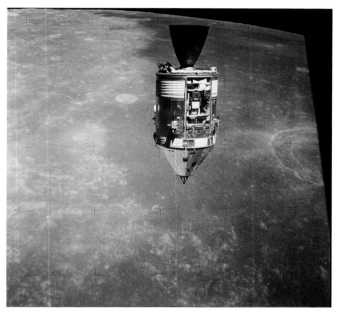

Figure 6.14   The Apollo command module in orbit around the Moon. (Photo: NASA)

Position and orbit

The Moon spins like a top on its axis as it travels around the Earth, and at the same time the Earth is orbiting the Sun. The Moon's orbit is slightly elliptical and its mean orbital radius is 384 400 kilometres. At closest approach (perigee) the orbital radius is 356 000 kilometres and at its greatest distance (apogee) it is 407 000 kilometres.

The Moon takes 27.3 days to go once around the Earth and it also takes this time to rotate once on its own axis. Because of this, we always see the same side of the Moon from Earth. As the Moon orbits the Earth, the angle between the Earth, Moon and Sun changes, and we see this as the cycle of Moon's phases (Figure 6.15). The time between successive new moons is actually 29.5 days, slightly different from the Moon's orbital period (measured against the stars), because the Earth moves a significant distance in its orbit around the Sun in that time.

The Moon appears to wobble on its axis because its orbit is elliptical. As a result we can see a few degrees of the far side of the Moon's surface from time to time. Most of the far side was completely unknown until the Soviet spacecraft **Luna 3** photographed it in 1959. The far side of the Moon gets sunlight half the time. Whenever we see less than a full moon, some sunlight is falling on the far side. Throughout each cycle of lunar phases, both sides of the Moon get equal amounts of sunlight.

Gravity keeps the Moon in orbit around the Earth and produces the tides. Tidal forces deform the oceans, causing them to rise at some places and to settle elsewhere. There are two areas of high tide on the Earth at any given time, one on the side closest to the Moon, the other on the

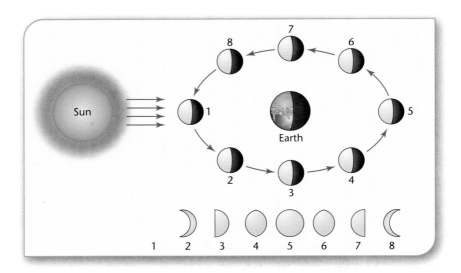

Figure 6.15   Phases of the Moon.

DID YOU KNOW?

About a century ago, Sir George Darwin (son of Charles Darwin) put forward the idea that the Moon is moving further away from the Earth. To test this theory, the Apollo 12 astronauts placed a set of reflectors on the Moon to measure the Moon's distance from Earth accurately. Pulses of strong laser light were fired at the Moon from Earth, and the time taken for the light to bounce off the reflectors and return to Earth was measured. Knowing the speed of light and the time for the pulses to return enabled astronomers to calculate the distance to an accuracy of less than a metre. From such measurements over time, astronomers have found that the Moon is spiralling away from the Earth at a rate of about 4 cm per year. The cause of this motion is thought to be tidal interactions between the Moon and Earth. Tidal forces are also causing the Earth's rotation to slow by about 1.5 milliseconds per century.

opposite side of the Earth. Low tides occur where the Moon is on the horizon.

The Sun also distorts the shape of the oceans, but only half as much as the Moon, because the Sun is nearly 400 times further away.

Density and composition

The Moon has a smaller mass, diameter and average density than Earth. Because of this, the strength of gravity on the Moon's surface is one-sixth that of Earth. A 75 kg person on Earth weighs 735 N, but on the Moon the same person would weigh only 122 N. The Moon's escape velocity is only 2.4 km/s.

Scientists have found out what the interior of the Moon is like from seismic (moonquake) evidence made at the Apollo landing sites. The Moon has a crust, mantle and core. Although these regions are similar to those of the Earth, the proportions are different. The Moon's crust averages 68 kilometres thick and the thickness varies from a few kilometres under Mare Crisium on the visible side, to 107 kilometres near the crater Korolev on the far side. The crust is thicker on the far side than on the side that faces Earth. This means the Moon is slightly egg-shaped, with the small end pointing towards Earth. As a result, the Moon's centre of mass is 2 kilometres closer to Earth than its geometric centre. The difference in thickness also helps explain why most of the mare basalt lavas are confined to the near side of the Moon. On the near side of the Moon the lavas would have reached the surface more easily.

Seismometers placed on the Moon by the Apollo astronauts found moonquakes occur on the Moon on a fairly regular cycle of about two weeks. They apparently result from the tidal stresses induced in the Moon as it rotates about Earth. Most moonquakes measure less than 3 on the Richter scale.

Figure 6.16   Instruments placed on the Moon by astronauts. (Photo: NASA)

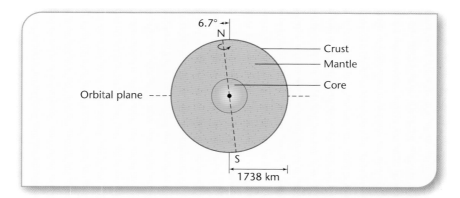

Figure 6.17 Internal structure of the Moon.

Below the crust is the Moon's mantle. The Moon's mantle probably makes up most of its interior. Unlike the Earth's mantle, however, the Moon's mantle is only partially molten.

The mantle is solid down to a depth of about 800 to 1000 km. The composition of the upper mantle may be deduced from the composition of the mare lavas, which came from these regions. Below about 1000 km the mantle becomes partially molten. Evidence for this came mainly from seismic data collected when a large meteorite weighing about 1 tonne hit the far side of the Moon in July 1972. At the centre of the Moon there may be a small iron-rich core perhaps only 800 km in diameter, but its existence is uncertain.

Analysis of Moon rocks shows no evidence for formation in a different part of the solar system from Earth. There is some evidence to support the theory that the Moon may have been once part of the Earth. The bulk density of the Moon is close to that of the silicate mantle of the Earth; however the bulk composition of the Moon is different from that of Earth's mantle. The Moon as a whole contains a higher proportion of iron and a lower proportion of magnesium than the Earth's mantle. The Moon also has a lower proportion of lead but a higher proportion of calcium, aluminium and uranium than Earth.

Information from the Moon rocks supports an 'impact theory' for the formation of the Moon. In this theory a large object, about the size of Mars, hit Earth in its first 100 million years of life. This collision ejected a lot of rock material from the Earth's surface into orbit, forming a 'debris ring'. This ring gradually condensed into the Moon. Such an impact could also have tipped the Earth off its axis and so created the seasons.

The surface

There are two main types of terrain on the Moon, the heavily cratered and very old highlands or terrae, and the relatively smooth and younger plains or maria (singular **mare**). These can be identified by the naked eye. Most of the surface is covered with regolith, a mixture of fine dust and rocky debris produced by meteor impacts. This layer ranges in thickness from 1 to 20 metres. Unlike Earth's soil, which has decayed biological matter in it, the Moon's regolith does not contain any biological matter.

The maria make up about 16 per cent of the Moon's surface and are sometimes called 'seas' even though they contain no water (mare means 'sea' in Latin). Maria are huge impact basins that have been covered by molten lava. Most maria exist on the side of the Moon facing Earth. The more prominent maria are Mare Tranquillitatis (Sea of Tranquillity), Mare Nubium (Sea of Clouds), Mare Nectaris (Sea of Nectar) and Mare Serenitatis (Sea of Serenity). The largest mare, Mare Imbrium (Sea of Showers), is circular and measures 1100 km in diameter. Like most maria, it is 2 to 5 km below the average lunar elevation.

Rocks brought back to Earth from the maria are solidified lava (mainly basalt), which suggests that the Moon's surface was once molten. These rocks have a composition similar to those found in volcanic rocks on Hawaii or Iceland – they contain heavy elements like iron, manganese and titanium. The molten lava has come from inside the Moon and has risen to the surface through large impact fractures in the crust.

Figure 6.18   Moon rock. (Photo: J. Wilkinson)

There is also some tectonic activity in the maria caused by the weight of basalts pushing on the crust. At the edges of the maria, the basalts are stretched, causing fracturing and faulting. In the interior of maria, the basalts are compressed, resulting in folding that produces wrinkle ridges. Most maria also contain small craters and occasional cracks (lava tunnels or channels) called rilles. In the highlands, tectonic activity has produced small scarps.

Lunar probes have shown that the far side of the Moon contains one prominent mare, Mare Moscoviense, and is heavily cratered. The cratered area on the far side is 4 to 5 km higher than the average lunar elevation.

The Moon's surface is covered with meteorite impact craters that vary in size from tiny pits to huge craters hundreds of kilometres in diameter. Virtually all the craters are round and the result of meteorite impact. Some of the craters have rays or streaks extending outwards from their centre, while others have raised peaks at their centre. These peaks occur because the impact compresses the crater floor so much that afterwards the crater rebounds and pushes the peak upwards. As the peak goes up, the crater walls collapse and form terraces. One of the most striking craters with a central peak is Copernicus. Copernicus crater is about 92 km across and 800 million years old. Rays are often formed when material is ejected and scattered across the surface during large impacts. The most striking crater with rays is Tycho, which formed about 109 million years ago.

Most of the craters on the near side are named after famous figures in the history of science, such as Tycho, Copernicus and Kepler. Other craters bear the names of philosophers such as Plato and Archimedes. Features on the far side have modern names such as Apollo, Gagarin and Korolev, with a distinctly Russian bias, since the first images of the far side were obtained by Luna 3. The largest impact basin (crater) is the Aitken basin, 2500 km wide and 12 km deep, at the south pole on the far side. The Imbrium basin, about 1800 km wide, and the Crisium basin, about 1100 km wide, are both found on the near side. The Orientale basin, 1300 km wide, located on the western limb, is a splendid example of a multi-ring crater.

DID YOU KNOW?

On 7 November 2005, NASA scientists observed an explosion on the Moon. The blast, equal in energy to about 70 kg of TNT, occurred near the edge of Mare Imbrium. The explosion was caused by a 12 cm-wide meteoroid slamming into the surface at about 27 km/s. Unlike Earth, the Moon has no atmosphere to burn meteoroids up, so they hit the ground and explode. The scientists captured the impact on video while observing through a 10-inch telescope. The impact gouged a crater in the Moon's surface about 3 m wide and 0.4 m deep.

In contrast to the dark maria, the light-coloured highlands are elevated regions that make up about 84 per cent of the lunar surface. Some of the highlands are mountains or ridges that are the rims of large basins formed from material uplifted after impacts. One of the highest mountains on the Moon, the Apennines, forms part of the Imbrium basin.

Highland rocks are light anorthosites (feldspars) rich in calcium and aluminium. Many highland rocks brought back to Earth are impact breccias, which are composites of different rocks fused together as a result of meteorite impacts.

Radioactive dating of Moon rocks has shown the mare rocks to be between 3.1 and 3.8 billion years old, while the highland rocks are between 4.0 and 4.3 billion years old.

None of the rocks contains evidence that water once existed on the Moon, and no traces of life have been found.

NASA's Lunar Reconnaissance Orbiter, scheduled to launch in October 2008, will release a probe that will probably crash into the lunar south pole, possibly kicking up a dust cloud that can be tested to see if it includes water vapour. India is also expected to launch a space probe in 2008, carrying a radar instrument that can distinguish between ice and rugged terrain.

Table 6.6 Percentage composition of Moon rocks

Mineral	Mare basalt	Highlands	Whole Moon
Silicon dioxide	45.0	45.0	44.0
Titanium dioxide	3.0	0.6	0.4
Aluminium oxide	8.5	25.0	6.0
Iron oxide	21.0	6.5	12.0
Magnesium oxide	12.0	8.5	33.0
Calcium oxide	10.0	14.0	4.5
Sodium oxide	0.2	0.4	0.10
Potassium oxide	0.06	0.07	0.01

Figure 6.19   The Moon's surface as seen by Apollo astronauts. (Photo: NASA)

Atmosphere

The Moon has no real atmosphere and no liquid water. Atmospheres are held in place by gravity, and the Moon has so little gravitational pull that it is unable to hold any of the gases such as those that make up the Earth's atmosphere.

Temperature

There is a large range of temperatures on the surface of the Moon because of its lack of atmosphere. On the night side temperatures can fall to

−184°C, while on the parts of the Moon facing the Sun temperatures can reach 130°C. At the poles, temperatures are constantly low, about −96°C. Some polar regions are in permanent shadow.

Because the Moon rotates once on its axis every 27.3 days, night and day at any point on the Moon last about fourteen Earth days. On the side of the Moon that always faces Earth, 'phases of the Earth' would be observed. Part of the long period of night would be 'Earth-lit', just as we have 'Moon-lit' nights on Earth.

Magnetic field

The Moon has no global magnetic field, but some rocks brought back by the Apollo astronauts exhibit permanent magnetism. This suggests that there may have been a global magnetic field early in the Moon's history.

With no atmosphere and no magnetic field, the Moon's surface is directly exposed to the solar wind. Since the Moon's early days, many charged particles from the solar wind would have become embedded in the Moon's regolith (surface material). Samples of regolith returned by the Apollo astronauts confirmed the presence of these charged particles.

WEB NOTES

For fact sheets on any of the planets, including Earth and the Moon, check out:
<http://nssdc.gsfc.nasa.gov/planetary/planetfact.html>
<http://apod.nasa.gov/apod/earth.html>

CHAPTER 7

MARS –
THE RED PLANET

MARS IS ONE of Earth's neighbours in space. Many people have considered Mars to be the most likely planet, apart from Earth, to contain life. Mars is the fourth planet from the Sun, orbiting at an average distance of

Figure 7.1   A view of Mars from the Mars Global Surveyor in 1997. (Photo: NASA)

228 million kilometres. This distance is about one and a half times the distance Earth is from the Sun. Radio signals take between 2.5 and 20 minutes to travel one way between Earth and Mars, depending on where Mars is in its orbit in relation to Earth. At times Mars is the third brightest planet in our night sky, after Venus and Jupiter. It has a diameter of 6794 km, about half Earth's diameter, making it the seventh largest planet. Mars is often referred to as the 'red planet' because it appears red from Earth. This colour comes from the large amounts of red dust that cover its surface. The planet is thought to have formed about 4.5 billion years ago, at the same time as the other planets in the solar system. Because it is relatively close to the Sun, Mars must have been hot and in a molten state before it cooled to become a solid planet.

EARLY VIEWS ABOUT MARS

Mars is named after the ancient Roman god of war. Both the ancient Greeks and Romans associated Mars with war because its colour resembles that of blood. The Greeks called the planet Ares. The two moons of Mars, Phobos (fear) and Deimos (panic), are named for the sons of the Greek god of war. The month of March derives its name from Mars.

Mars has been known since prehistoric times and many ancient astronomers have studied its motion in the night sky. The Danish astronomer Tycho Brahe (1546–1601) made decades of observations of Mars's motion. Brahe's work was continued by Johannes Kepler (1571–1630), who used the observations to develop the first two of his three laws of planetary motion. These included the conclusion that planets orbit in an elliptical path, with the Sun at one focus.

Another early astronomer to study Mars was the Italian–French Giovanni Domenico Cassini, who, in 1666, made the first reasonably accurate measurements of Mars's axial rotation period, which he found to be 37.5 minutes longer than that of the Earth. Cassini was also the first to observe the Martian polar ice-caps. The first observations of the surface markings were made in 1659 by Christiaan Huygens, who drew the dark triangular feature we now know as the large plateau Syrtis Major. Huygens also estimated the length of the Martian day to be about 24 hours.

In 1777, William Herschel measured the tilt of the axis of Mars and deduced it must have seasons like Earth because it underwent regular changes in its polar ice-caps.

Wilhelm Beer and Johann von Madler made the first detailed maps of the Martian surface features during the 1830s.

In 1877 the Italian astronomer Giovanni Schiaparelli reported seeing canals on the surface of Mars. American Percival Lowell reported further canals. Lowell thought Mars was a desert world and that inhabitants of Mars used the canals to carry water from the ice-caps to equatorial regions. Exploration of Mars by space probes has since disproved the existence of such canals.

PROBING MARS

People on Earth have observed Mars through telescopes based on Earth and in space. Early space probes carried telescopes and cameras to observe

Table 7.1 Details of Mars

Distance from Sun	227 940 000 km (1.52 AU)
Diameter	6794 km
Mass	6.42×10^{23} kg (0.107 times Earth's mass)
Density	3.95 g/cm³ or 3950 kg/m³
Orbital eccentricity	0.093
Period of revolution (length of year)	687 Earth days or 1.881 Earth years
Rotation period	1.029 Earth days
Orbital velocity	86 868 km/h
Tilt of axis	25.2°
Average temperature	−60°C
Number of moons	2
Atmosphere	Carbon dioxide
Strength of gravity	3.6 N/kg at surface

Mars as they flew past it. Later probes went into orbit around Mars and returned much more data. More recently, probes have successfully landed on the Martian surface, but to date no human has yet set foot on Mars.

The first three space probes directed at Mars were launched by the USSR between 1960 and 1962, but failed to leave Earth orbit. **Mars 1**, launched by the USSR on 1 November 1962, was the first probe to fly past the planet, but it failed to return data. The USA attempted a fly-by of Mars in November 1964 with **Mariner 3**, but its solar panels did not open and it is now in solar orbit. **Mariner 4** arrived at Mars on 14 July 1965 and passed within 9920 kilometres of the planet's surface, sending back 22 black and white photographs of a barren, cratered surface. The thin atmosphere was confirmed to be composed of carbon dioxide, and a weak magnetic field was detected.

Mariner 6 arrived at Mars on 24 February 1969 and passed within 3437 kilometres of the planet's equator. **Mariner 7** arrived on 5 August 1969 and passed within 3551 kilometres of the equator. The probes measured the atmospheric temperature and pressure, and surface composition. Together, the two probes took over 200 photographs.

The **Mars 2** space probe reached Mars in 1971 and released a lander that crashed into the Martian surface when its rockets failed to slow it down. No data were returned, but it was the first human-made object to be placed on Mars. The Mars 2 orbiter returned data until 1972. **Mars 3** arrived at Mars on 2 December 1971 and a lander was successfully placed on the surface, however it returned video data for only 20 seconds.

The first US spacecraft to enter an orbit around another planet was **Mariner 9**, on 3 November 1971. At the time of its arrival a huge dust storm was in progress on Mars and many experiments had to be delayed until the storm had finished. The probe sent back the first high-resolution images of the moons, Phobos and Deimos. Mariner 9 took over 7000 images of Mars and showed that giant volcanoes and river-like features exist on the surface. The probe is still in orbit around Mars.

Of the four separate probes launched by the USSR in 1973, **Mars 5** was the most successful, going into orbit around Mars on 12 February 1974. It returned 70 high quality images to Earth.

Viking 1 went into orbit around Mars on 19 June 1976, and its lander touched down on 20 July. **Viking 2** went into orbit on 24 July 1976, and its lander reached the surface on the opposite side of Mars on 7 August 1976. Both landers sent back a great deal of information about surface features and atmospheric conditions as well as conducting experiments to search for micro-organisms. No conclusive evidence of life was found. Pictures showed the Martian sky to be reddish as a result of the fine dust suspended in the atmosphere. The atmospheric pressure was found to be less than 1 per cent of that at the Earth's surface. The orbiters mapped the entire planet's surface, acquiring over 52 000 images.

Missions to Mars in the late 1980s were generally unsuccessful. In 1988 the USSR launched **Phobos 1** and **2** to study the moons of Mars. Contact with Phobos 1 was lost en route to Mars as a result of a command error.

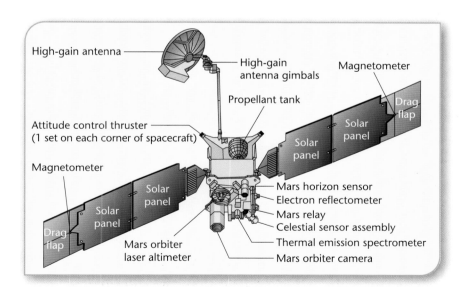

Figure 7.2   Mars Global Surveyor.

Phobos 2 entered orbit around Mars on 30 January 1989, and came to within 120 kilometres of the moon before it failed.

The **Mars Observer** probe launched in 1992 by NASA was intended to provide the most detailed images of Mars ever obtained, but communications with it were lost just as it was about to enter orbit around the planet in August 1993.

The **Mars Global Surveyor** was the first mission of a new decade-long program of spacecraft to be sent to Mars. Launched on 7 November 1996, and consisting of an orbiter and robotic lander, the probe was designed to orbit Mars over a two-year period and collect data on the surface composition and features, atmospheric dynamics and magnetic field. The probe was inserted into a low altitude, nearly polar orbit on 12 September 1997 and it now circles Mars once every two hours. The mission has studied the entire Martian surface, atmosphere and interior, and has returned more data about the red planet than all other missions combined.

Mars Pathfinder was the first completed mission in NASA's Discovery Program of low-cost, rapidly developed planetary missions with specific scientific goals. Pathfinder arrived at Mars on 4 July 1997. It used an innovative method of directly entering the atmosphere, assisted by a parachute to slow its descent and a giant system of airbags to cushion the impact. The landing site was an ancient, rocky, flood plain in Mars's

Figure 7.3   The surface of Mars from Pathfinder Rover in 1997. (Photo: NASA)

northern hemisphere, known as Ares Vallis. Scientists believe a large volume of water cut the flood plain in a short period of time. A six-wheel robotic rover named Sojourner rolled onto the Martian surface on 6 July. Mars Pathfinder returned 2.6 billion bits of data, including more than 16 000 images from the lander and 550 images from the rover, as well as more than 15 chemical analyses of rocks and extensive data on winds and

Figure 7.4   Mars Pathfinder rover Sojourner and the rock 'Yogi'. (Photo: NASA)

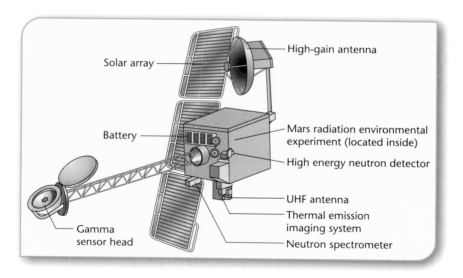

Figure 7.5   The Mars Odyssey space probe.

other weather factors. The last transmission of data was on 27 September 1997, 83 days after the mission landed on the surface.

In April 2001, the USA launched the **Mars Odyssey** space probe. The probe carried instruments to analyse the chemical composition of the surface and rocks just below the surface. The mission also looked for the presence of water ice on Mars and looked for radiation hazards in the space surrounding Mars. Mars Odyssey went into orbit around Mars in October 2001.

In June and July of 2003, NASA launched two space probes containing robotic rovers, Spirit and Opportunity, as part of the **Mars Exploration Rover** project. Spirit was launched on 10 June 2003 while Opportunity was launched on 7 July.

Spirit landed on 4 January 2004, and Opportunity landed on 25 January. Spirit landed in a plain covered with volcanic rocks and small craters. It found evidence that small amounts of water may have once existed in the cracks of rocks. Spirit spent 10 weeks exploring its landing area, travelling

Figure 7.6 Both Mars exploration rovers are solar-powered and have six wheels. They are about the size of a golf cart. The rovers carry panoramic cameras, spectrometers for mineral analysis, a rock abrasion tool, and a microscopic imager. (Photo: NASA)

more than 3 km across the Martian soil. Opportunity landed close to a thin outcrop of rocks in a small crater called Eagle. Within two months, it had found evidence in those rocks that a body of salty water deep enough to splash in once flowed through the area. This suggested that the past environment could have been hospitable to life. Opportunity climbed out of Eagle crater on 22 March 2004 and headed towards another crater called Endurance. It set a one-day record of 140.9 metres on 17 April and reached the rim of the crater on 30 April. Instruments on the rover found that rocks in this new area had also been exposed to water in the past. The rovers, however, were not equipped to detect life or fossil remains.

The two rovers survived the effects of a large Martian dust storm during 2007, which darkened the Martian sky by 99 per cent and reduced available power from their solar arrays to a bare minimum. On 13 September 2007, Opportunity commenced a descent into the 800-metre-wide Victoria crater, where it encounted older rocks than those on the surface. Spirit had reached 'Home Plate', a plateau of layered bedrock that has planetary scientists excited since it exhibits plenty of evidence of past water. In March 2007, Spirit discovered a patch of bright-toned soil that was found to contain 90 per cent silica. Scientists believe this patch could have been concentrated by water.

By January 2008, the two rovers had been operating on Mars for four years, fifteen times their planned life-time. The rovers are now expected to last till at least 2009.

The ESA launched the **Mars Express** probe on 2 June 2003. The two-year mission was so-called because of its rapid and streamlined development time. The probe was designed to answer questions about the geology, surface environment, water history and potential for life on Mars. The probe went into orbit around Mars during 2005 and its mission has been extended to at least 2009.

NASA's **Mars Reconnaissance Orbiter** arrived at the red planet in March 2006 and spent half a year gradually adjusting the shape of its orbit. During 2007, the probe orbited the planet once every 24 hours and returned data to Earth at a rate faster than any previous mission to Mars (6 megabits per second, fast enough to fill a CD-ROM every 16 minutes). The orbiter is examining Mars in unprecedented detail, including water and mineral distribution, features and future landing sites. So far the probe has discovered channels in a fossil delta, troughs in sand dunes and hardware from the landing of the rover Opportunity. The mineral-mapping instrument on the orbiter has also been put into operation. There is also evidence that liquid or gas has flowed through cracks in underground rocks.

A US space probe named Phoenix Mars Lander was launched from Cape Canaveral, Florida, on 4 August 2007. The probe landed on Mars on 25 May 2008 with the aid of retrorockets. The probe is digging for clues to past and present life in the northern polar region. The solar-powered craft is equipped with a 2.35-metre robotic arm that will enter vertically into the soil and lift soil samples to two instruments on the deck of the lander. One instrument will check for water and carbon-based chemicals, considered essential for life, while the other will analyse the soil chemistry. Phoenix is experiencing Martian temperatures that range from −73°C to −33°C.

Table 7.2 Significant space probes to Mars

Probe	Country of origin	Launched	Comments
Mars 1960A	USSR	1960	Failed to reach Earth orbit
Mars 1960B	USSR	1960	Failed to reach Earth orbit
Mars 1962A	USSR	1962	Failed after final rocket stage exploded
Mars 1	USSR	1962	Communications failed en route
Mars 1962B	USSR	1962	Failed to leave Earth orbit
Mariner 3	USA	1964	Mars fly-by attempt, solar panels failed
Mariner 4	USA	1964	Passed by Mars in 1965, returned 22 close-up photographs
Zond 2	USSR	1964	Contact lost en route to Mars
Mariner 6	USA	1969	Passed within 3437 km of surface, many photographs taken
Mariner 7	USA	1969	Passed within 3551 km of surface, photos taken
Mariner 8	USA	1971	Failed to reach Earth orbit
Kosmos 419	USSR	1971	Failed to leave Earth orbit

Probe	Country of origin	Launched	Comments
Mars 2	USSR	1971	A lander was released from the orbiter and crashed on surface
Mars 3	USSR	1971	First successful landing on Mars; relayed 20 seconds of data
Mariner 9	USA	1971–2	Entered orbit around Mars, took photographs of the two moons
Mars 4	USSR	1973	Flew past Mars as a result of engine malfunction; returned images
Mars 5	USSR	1973	Orbited Mars in 1974, 70 images taken
Mars 6	USSR	1973	Entered orbit, lander failed on descent, some data returned
Mars 7	USSR	1973	Failed to enter orbit, lander missed planet
Viking 1	USA	1975	Both orbiter and lander successful at photographing Mars
Viking 2	USA	1975	Orbiter and lander successful; found no conclusive signs of life
Phobos 1	USSR	1988	Lost en route to Mars
Phobos 2	USSR	1988	Orbited Mars, came close to Phobos, but failed
Mars Observer	USA	1992	Contact lost as it was entering orbit around Mars
Mars Global Surveyor	USA	1996	Mapped entire surface in detail
Mars Pathfinder	USA	1996	Robotic rover successfully used to explore surface
Mars Climate Orbiter	USA	1998	Reached Mars but missed target altitude
Mars Polar Lander	USA	1999	Communication lost on atmospheric entry
Deep Space 2	USA	1999	Contact lost as the two probes neared Mars
Mars Odyssey	USA	2001	Orbited Mars for three years
Mars Exploration Rovers	USA	2003	Rovers Spirit and Opportunity used to explore surface of Mars
Mars Express	ESA	2003	In orbit around Mars during 2005
Mars Reconnaissance Orbiter	USA	2005	Multipurpose spacecraft to orbit Mars
Phoenix Mars Lander	USA	2007	Landed near Mars's north pole, 25 May 2008

POSITION AND ORBIT

Mars orbits the Sun in an elliptical orbit that has the third highest eccentricity of all the planets' orbits. Its mean distance from the Sun is about 228 million kilometres, placing it about one and a half times further from the Sun than is Earth. At perihelion, Mars is 208 million kilometres from the Sun, while at aphelion it is 249 million kilometres. The difference between the two is about 46 million kilometres, whereas for Earth the difference between perihelion and aphelion is only 5 million kilometres. This has a major influence on Mars's climate and results in a wide range of seasonal temperatures.

Mars orbits the Sun with a velocity of about 86 868 km/h and takes 687 Earth days to complete the trip. The planet takes 24.6 hours (1.029 Earth days) to rotate once on its axis, which is tilted at an angle of 25.2° to the vertical.

About every 780 days Mars passes through a point in its orbit where from Earth it appears opposite the Sun in the sky (opposition). Because of

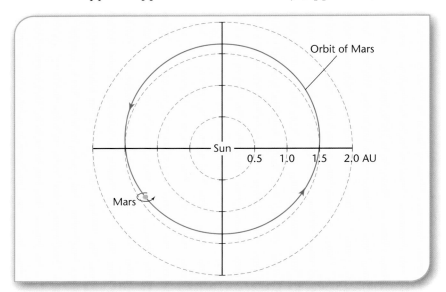

Figure 7.7 Orbit of Mars around the Sun (distance circles are in astronomical units, AU)

its eccentric orbit, Mars's distance at opposition varies, so its apparent size and brightness also change. The most favourable opposition for observation is when Mars is closest to both the Earth and the Sun (this occurs about once every 17 years). Future oppositions for Mars are 29 January 2010, 3 March 2012, 8 April 2014, and 22 May 2016.

Mars can be studied easily from Earth using a telescope of moderate power. Dark markings and the white polar ice-caps may be seen on the surface, depending on the distance of Mars from Earth.

DENSITY AND COMPOSITION

Even though Mars is more than half the diameter of Earth, it has only about one-tenth of its mass. Detailed knowledge about the interior of Mars is limited because of the lack of seismic data.

The average density of Mars is the lowest of the terrestrial planets (3.95 g/cm^3 compared with Earth's 5.52 g/cm^3). This suggests the iron-bearing core of Mars is smaller than Earth's core. In fact the core is thought to have a radius of only 1100 km (but some estimates have it as high as 2000 km). The core makes up only about 6 per cent of the planet's mass,

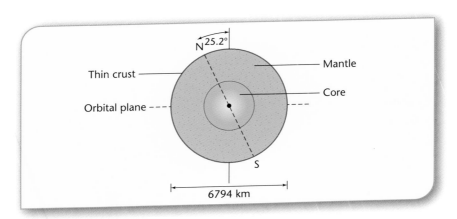

Figure 7.8   The interior structure of Mars.

145

compared to Earth's core, which makes up about 32 per cent of its mass. A weak magnetic field suggests the core is no longer liquid or that currents within it are slow. Surrounding the core is a molten rocky mantle about 2200 km thick that is less dense than the core. The outer crust of the planet varies in thickness from about 20 km to 150 km.

The strength of gravity on Mars is about a third of that on Earth. A 75 kg person weighs 735 N on Earth, but would weigh only 270 N on Mars.

THE SURFACE

Although Mars is much smaller than Earth, its surface area is about the same as the land surface area of Earth. The Viking landers provided amazing views of the surface of Mars. Our first view of the surface was obtained from the Viking 1 lander in July 1976. Pictures revealed a rocky, desert-like terrain. Two weeks later, Viking 2 set down on the opposite side of Mars, and images returned to Earth showed the same kind of rocky surface as revealed by Viking 1. Both landing sites contained rocks ranging from pebbles to boulders, in an orange-red, fine-grained soil.

Much of the Martian surface is very old and cratered, but there are also much younger rift valleys, ridges, hills and plains.

There is highly varied terrain on Mars, with highlands in the southern hemisphere and lowlands in the northern hemisphere. The highlands are the oldest terrain (about 3 billion years old), and many parts are heavily cratered. The oldest terrain also contains small channels that may have been carved by flowing water. Smooth plains between the cratered areas are volcanic in origin.

The lowlands in the north contain mostly plains with few craters, indicating they probably formed after the period of bombardment by meteorites. The region between the highlands and lowlands is marked by an escarpment or long cliff. The reason for this abrupt elevation change could be a very large impact during Mars's past. A three-dimensional map of Mars that clearly shows these features was produced by the Mars Global Surveyor space probe.

The planet's western hemisphere contains a distinct bulge about 10 km high and 8000 km long, called the **Tharsis Rise**. This region contains the greatest concentration of volcanic and tectonic activity on Mars. Many volcanoes, fractures and ridges, and the enormous **Valles Marineris** canyon system, are linked to this rise. Valles Marineris was named after the Mariner 9 probe that discovered it. The canyon is about 8 km deep and 4500 km long (the same as the distance between Sydney and Perth). Smaller tributary canyons are as large as the Grand Canyon on Earth. Valles Marineris probably formed from rifting, or the pulling apart of the Martian crust, at the same time as the Tharsis Rise formed.

Mars also contains the largest volcanic mountain in the solar system, **Olympus Mons**. This mountain rises to a height of 24 km above the surrounding plains, is more than 500 km wide, and is rimmed by a cliff

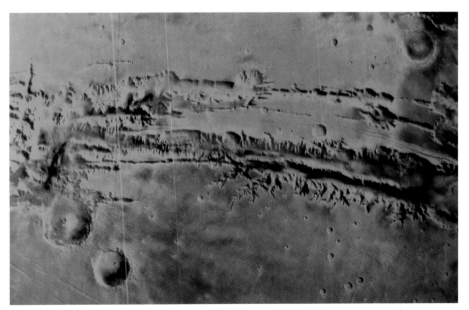

Figure 7.9 Valles Marineris extends about a quarter of the way around Mars. (Photo: NASA)

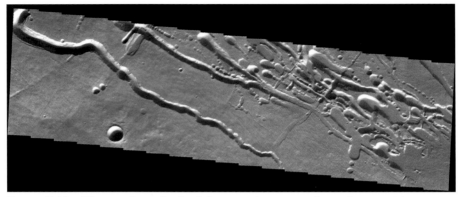

Figure 7.10   The southern flank of the giant Martian volcano Pavonis Mons contains channels, grooves and pits (photo taken by Mars Odyssey).

6 km high. The volcano's summit has collapsed to form a volcanic crater or caldera about 90 km across.

There are three other prominent volcanoes on Mars: **Arsia Mons**, **Pavonis Mons** and **Ascraeus Mons**. Each is over 20 km high and forms part of a volcanic chain near the centre of Tharsis Rise. **Alba Patera** is a low-relief volcano about 2 km high and 700 km across situated near the rise's northern edge. Most of the volcanoes on Mars are in the northern hemisphere, while most of the impact craters are in the southern hemisphere.

The Martian surface contains many large basins (impact craters) formed when large meteors hit the surface. The largest basins formed by impacts are **Hellas** (with a diameter of about 2000 km), **Isidis** (1900 km) and **Argyre** (1200 km).

The soil on Mars was analysed by the Viking landers and found to be slightly magnetic, indicating it contains iron. Further analysis showed the rocks at both landing sites were rich in iron, silicon and sulfur. As a result the Martian soil is described as an iron-rich clay. The soil is also rich in chemicals that effervesce (fizz) when moistened. Unstable chemicals called peroxides exist in the soil, and these break down in the presence of water to release oxygen gas. **Mars Pathfinder** was able to identify the presence of

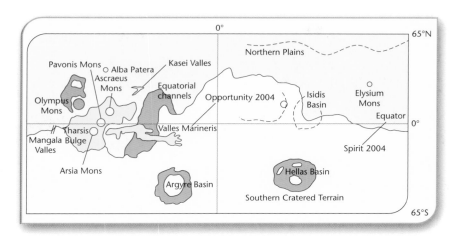

Figure 7.11 Simplified map of geological features on Mars. The shaded area is the Northern Plains. The lower area is the southern cratered terrain.

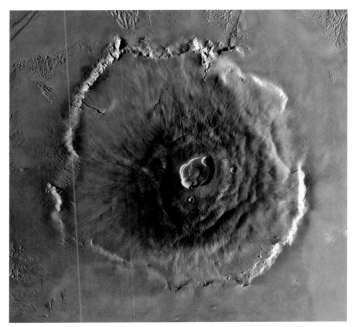

Figure 7.12 Olympus Mons is the largest mountain in the solar system. (Photo: NASA)

conglomerates like those that are formed by running water on Earth. This evidence suggests that Mars might have had a warmer past in which liquid water was stable.

Like Mercury and the Moon, Mars appears to lack active plate tectonics at present, since there is no evidence of recent horizontal motion of the surface. Although there is no current volcanic activity, the Global Surveyor space probe showed that Mars might have had tectonic activity in its early history. Mars does not appear to have a crust made up of several large plates, like Earth has – it may be a single-plate planet.

Mars Global Surveyor found many landforms on Mars seem to have been formed or altered by running water. The channels that form valleys in the cratered highlands are similar to those formed by water on Earth. Some eroded valleys are huge, for example the **Kasei Valles** cuts over a kilometre deep into the volcanic plains of the Tharsis Rise, and is over 2000 kilometres long. It is thought that some of the water flows in the past were huge in volume and occurred when internal heat or meteorite

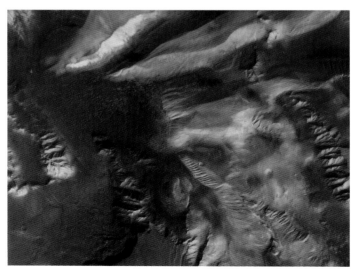

Figure 7.13 Mangala Valles lies on the boundary between the planet's heavily cratered terrain and the volcanic plains. The area contains a number of outflow channels, the result of flooding millions of years ago. (Photo: NASA)

impacts released groundwater in sudden floods. Such flows were brief, however, since Mars does not have enough water to sustain continuous flow. Some scientists believe that many of the geological features that appear to have been caused by water may have instead been formed by pyroclastic flows similar to those that occur during a volcanic eruption.

If there is evidence of water in Mars's past, where is it today? Most of the water is believed to exist as permafrost in the northern lowlands and maybe underground in the heavily fractured and cratered highlands, and as ice at the poles. In December 2006, scientists comparing photographs taken in 1999 and 2005 from the Mars Global Surveyor's orbiting camera discovered that water had flowed down the walls of a crater during this period.

Mars has two prominent polar ice-caps, which can be seen through telescopes from Earth. The two polar regions of Mars are mostly covered with layered deposits of solid carbon dioxide (dry ice), with some dust and

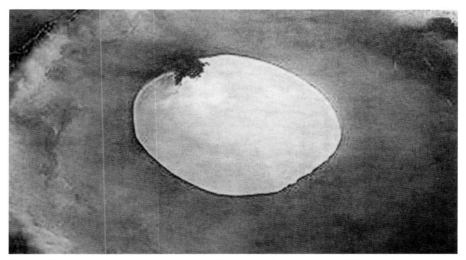

Figure 7.14 A lake of water ice was discovered on Mars in 2005 by the European Space Agency's probe Mars Express. The ice rises 200 metres above the floor of an impact crater located on Vastitas Borealis, a broad plain that covers much of the planet's far northern latitudes. The crater is 35 km wide and has a maximum depth of 2 km. The water ice appears to be present all year round. Atmospheric pressure and temperature around the icy patch are such that the ice cannot either melt into liquid water or sublime into water vapour. (Photo: European Space Agency)

water ice. The mechanism responsible for layering is thought to be climatic changes. The Viking landers found that seasonal changes in the extent of the polar ice-caps change the atmospheric pressure by about 25 per cent.

During summer in the northern polar region, carbon dioxide returns to the atmosphere, leaving a cap of water ice. The southern polar region reduces its size during summer but stays as frozen carbon dioxide.

Life may exist in the permafrost or under the polar ice-caps of Mars. On Earth, algae, bacteria and fungi have been found living in ice-covered lakes in Antarctica, and so future explorations of Mars's ice may reveal life-forms.

THE MARTIAN ATMOSPHERE

The Martian atmosphere is very different from Earth's, but there are some similarities. Mars has a very thin atmosphere composed of 95.3 per cent carbon dioxide, 2.7 per cent nitrogen, 1.6 per cent argon, and less than 0.2 per cent oxygen. The atmosphere is near its saturation point with water vapour (0.03%). In the past the Martian atmosphere may have been denser, but it is now one-hundredth the density of Earth's atmosphere. This low value is partly a consequence of the low gravitation field. Data returned by the Phobos space probe suggested that the solar wind is carrying away the weakly held atmosphere at a rate of 45 000 tonnes per year.

The average air pressure on Mars is about seven-thousandths of Earth's, but it varies with altitude from almost nine-thousandths in the deepest basins to about one-thousandth at the top of Olympus Mons.

The minute traces of water vapour can at times form clouds, particularly in equatorial regions around midday. Early morning fogs also appear in canyons and basins. Temperatures around the poles are often low enough for carbon dioxide to form a thin layer of cloud.

The atmosphere is thick enough to support strong winds and dust storms that sometimes cover large areas of the planet. At times, the dust storms can hide surface features from Earth view. Such storms occur most often during the southern hemisphere's spring and summer. In 1971, Mariner 9's view of Mars was obscured by a dust storm that lasted for two weeks. In 1977, thirty-five dust storms were observed, and two of these developed into global storms. Global storms spread rapidly, eventually enshrouding the whole planet in a haze that can last a few months.

Mars's thin atmosphere produces a greenhouse effect, but it is only enough to raise the surface temperature by 5°, which is much less than increases on Earth and Venus.

TEMPERATURE AND SEASONS

The two Viking landers functioned as weather stations for two full Martian years. Their data, together with information from the orbiters, has given us a good picture of the weather on Mars.

Mars has a greater average distance from the Sun than does Earth, and because of this it has a lower average surface temperature (–60°C). At perihelion, Mars receives about 45 per cent more solar radiation than at aphelion. As a result there is a large variation in surface temperatures during the Martian year. The lowest temperatures, of –125°C, occur in winter at the south pole; this temperature is the freezing point of carbon dioxide. The highest temperatures, of around 22°C, occur during summer in southern mid-latitudes. The large difference between equatorial temperatures and polar temperatures produces brisk westerly winds and low pressure systems, similar to cyclonic systems on Earth.

Mars is tilted on its axis at 25.2°, which is similar to Earth's tilt of 23.5°, and so it experiences four seasons. Each season lasts about twice as long as Earth's because Mars's orbit is much larger.

In Mars's northern hemisphere, spring and summer are characterised by a clear atmosphere with little dust. White clouds may be seen at sunrise near the horizon and at higher elevations. During winter, falling temperatures around the northern polar ice-cap cause carbon dioxide from the atmosphere to condense to renew the ice-cap. The carbon dioxide ice comes and goes with the seasons, but a permanent ice-cap of water ice remains.

The southern hemisphere summer occurs when Mars is closest to the Sun, and so southern summers are hotter than northern summers and winters are colder. During summer the southern polar ice-cap shrinks but a core of water ice remains.

MAGNETIC FIELD

Data from Mariner 4 indicated that Mars has a weak magnetic field. Unlike Earth, which has a single global magnetic field, Mars appears to have a

DID YOU KNOW?

The **Viking 1** and **2** landers were designed to search for life on Mars. Soil samples were analysed for organic compounds (containing carbon) that might indicate the presence of life. The Viking landers took soil samples from various places but no organic compounds were found. It is thought that some life-forms such as bacteria, algae and fungi may still exist under the permafrost or under the ice-caps. Future space missions will investigate this further.

The two Mars Exploration rovers **Spirit** and **Opportunity** found some evidence suggesting the past environment could have been hospitable to life, however these rovers were not equipped to detect life.

A meteorite discovered on Earth in 1984 originated from Mars. The meteorite, known as **ALH84001**, is believed to have been blasted off Mars by a comet or asteroid impact, since it contains gases and isotopes matching the atmosphere of Mars.

number of separate sources of magnetic force. The Mars Global Surveyor found that the strength of the magnetic field varies around the planet. The weak magnetic field suggests that the core of Mars is no longer liquid or that currents in the core are slow.

MARTIAN MOONS

Mars has two small moons that were discovered in 1877 by American astronomer Asaph Hall. Hall named them **Phobos** and **Deimos**. Phobos is the larger and innermost of the two moons, being 21 km in diameter. It takes 7 hours 39 minutes to orbit Mars, at an orbital radius of 9380 km. Deimos is only 12 km in diameter and takes 30 hours 18 minutes to orbit Mars, at an orbital radius of 23 460 km. Both moons have an irregular shape, low density, and many craters. They are hard to see from Earth because they are so small and reflect little light. The densities of each moon are so low that they cannot be pure rock. Both moons orbit in nearly circular orbits.

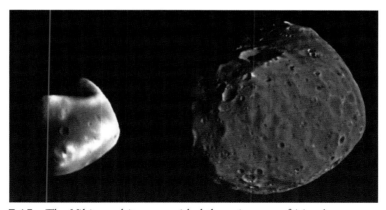

Figure 7.15 The Viking orbiters provided these images of Mars's moons. Phobos, the larger of the two, is irregularly shaped and measures roughly 28 km × 23 km × 20 km. Deimos is less cratered than Phobos and measures roughly 16 km × 12 km × 10 km. (Photo: NASA)

Table 7.3 Details of the moons of Mars

Name	Distance from Mars (km)	Period (days)	Diameter (km)	Discovered
Phobos	9380	0.32	21	1877
Deimos	23 460	1.26	12	1877

New images from **Mars Global Surveyor** indicate Phobos is covered with a layer of fine dust about a metre thick. The Soviet spacecraft **Phobos 2** detected a faint but steady out-gassing from Phobos. Unfortunately, Phobos 2 failed before it could determine the nature of the material.

Phobos is slowly being pulled closer to Mars (1.8 metres per century), and in about 50 million years it will either crash into the surface of Mars or break up into a ring. Deimos, on the other hand, appears to be getting further from Mars, slowing down as it does so.

Phobos was photographed by **Mariner 9** in 1971, **Viking 1** in 1977, and **Phobos 2** in 1989. This moon has an irregular shape and always has the

Figure 7.16   Phobos. Grooves and streaks on the surface were probably formed by ancient impacts. (Photo: NASA)

same face turned towards Mars. If you were standing on Mars's equator you would see Phobos rising in the west, move across the sky in only five and a half hours, and set in the east, usually twice a day. Deimos takes about two and a half days to cross the Martian sky.

Both moons are heavily cratered. Phobos has one large crater called Stickney (named after Angelina Stickney, Hall's wife) that is 10 km wide. The grooves and streaks on the surface of Phobos were probably caused by the Stickney impact. Deimos is less cratered than Phobos. The largest crater on Deimos is only 2.3 km across.

It is thought that these two moons may have been asteroids captured by Mars's gravitational field. Evidence to support the asteroid theory is that both moons reflect very little of the light that falls on them and they are very light for their size. The cratering on each moon suggests their surfaces are equally old – about three billion years. They are similar to C-type asteroids, which belong to the outermost part of the asteroid belt.

WEB NOTES

For information about Mars and the various space missions check out:
<http://marsprogram.jpl.nasa.gov>
<http://marsrovers.jpl.nasa.gov>
<http://www.space.com/mars/>
For fact sheets on any of the planets, including Mars, check out:
<http://nssdc.gsfc.nasa.gov/planetary/planetfact.html>

CHAPTER 8

THE ASTEROID BELT

THE **ASTEROID BELT** is a region between the orbits of Mars and Jupiter that contains over a million bodies, called **asteroids**. Asteroids are rocky and metallic objects that do not fit the definition of planets because they have not cleared the neighbourhood of their orbits. Most of these bodies are only a few kilometres in size and have irregular shapes because they are too small for their gravity to make them spherical (another requirement of the definition of a planet). One exception is Ceres, which is spherical and is classed as a dwarf planet. The asteroids all orbit the Sun in much the

Figure 8.1   The asteroid Ida as seen by the Galileo space probe in 1993. Ida has a length of 60 km and spins once every 4 hours 36 minutes. (Photo: NASA)

same direction. The belt contains about 200 asteroids larger than 100 kilometres in diameter and about 750 000 up to one kilometre across. There may be millions of even smaller asteroids.

Some asteroids have very elliptical orbits that cross Earth's orbit. These Earth-crossing asteroids have many craters on their surface because of impacts with other smaller bodies. Fragments of rock or iron ejected from the asteroids following impacts create bodies called **meteoroids**. A meteoroid entering Earth's atmosphere often heats up and glows, and appears like a like a shooting star across our night sky (called a meteor). Most meteoroids are about as old as the solar system itself.

The largest and first-known asteroid, **Ceres**, is about 950 km in diameter. It contains about one-third the total mass of all the asteroids. In 2006, Ceres was classified as a dwarf planet because it orbits the Sun, has enough mass to form a spherical shape, has not cleared the area around its orbit,

Figure 8.2 Fragments from asteroids sometimes enter Earth's atmosphere and are seen as meteors. Meteors that hit the surface can cause a crater such as this one at Wolfe Creek, Western Australia. (Photo: J. Wilkinson)

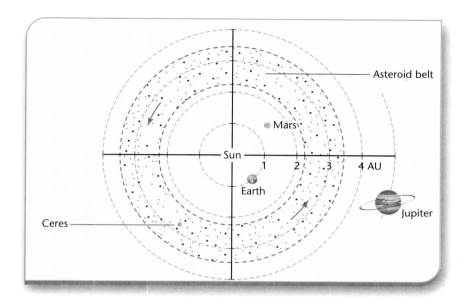

Figure 8.3 The asteroid belt is located between the orbits of Mars and Jupiter (distances circles are in astronomical units, AU).

and is not a satellite. Ceres is thus a dwarf planet within the asteroid belt. The second largest asteroid is **Vesta**, with a diameter of around 550 km, closely followed by Pallas, an irregularly shaped object about 540 km across. One of the smallest asteroids identified is **1991BA**, discovered in 1991, and only 6 m across.

EARLY VIEWS ABOUT THE ASTEROIDS

The word 'asteroid' means 'star-like'. This name probably arose because, viewed through a small telescope from Earth, asteroids look like points of light.

Ancient observers on Earth did not know the asteroids because they cannot be seen with the unaided eye. The idea that a planet-like body might exist between the orbits of Mars and Jupiter was suggested by Johann

Elert Bode in 1768. In January 1801, the Sicilian astronomer Giuseppi Piazzi discovered a body in a position similar to that predicted by Bode. This body was called **Ceres** in honour of the Roman goddess of plants and harvest, and was the first asteroid to be discovered. In March 1802, the German astronomer Heinrich Olbers discovered another faint asteroid that he called **Pallas** (after the Greek goddess of wisdom). Two more asteroids discovered in the early 1800s were called **Juno** and **Vesta**.

By 1850, 10 asteroids were known to orbit at average distances from the Sun of between 2.2 and 3.2 AU. These early findings were made visually by astronomers who spent many hours at telescopes, observing changes in positions of celestial objects against the background of stars. In 1891, the German astronomer Max Wolf made the first photographic discovery of an asteroid. The object was named **Brucia**, and was the 323rd asteroid to be found. By 1923, the list of asteroids had grown to over a thousand.

During the early years of discovery, mythological names were given to the asteroids. All the early names were female. This naming scheme was later abandoned. Some asteroids are named after countries, for example, 1125 China, while others are named after people, for example, 2001 Einstein. Permanent numbers are assigned to asteroids once their orbits have been calculated and confirmed.

Table 8.1 Details about the largest asteroid Ceres (a dwarf planet)

Distance from Sun	413 700 000 km (2.76 AU)
Diameter	950 km in diameter
Mass	9.46×10^{20} kg
Density	2.07 g/cm^3 or 2070 kg/m^3
Orbital eccentricity	0.08
Period of revolution (length of year)	1681 Earth days or 4.60 Earth years
Rotation period	9.07 hours
Orbital velocity	17.9 km/s
Axial tilt	3°
Average temperature	−106°C
Atmosphere	Tenuous, may have some water vapour
Strength of gravity	0.27 N/kg

Astronomers are not certain how the asteroids originated, but many believe they are part of the solar nebula that failed to form a planet because of the strong gravitational pull of the planet Jupiter. Others believe the asteroids are the remains of a planet that was pulled apart. This is unlikely, since if all the asteroids were combined into a single planet it would have a diameter of only 1500 km. Thus it is more likely that the asteroids are the remains of rocky bodies that have survived from the early solar system.

PROBING THE ASTEROIDS

In October 1991, the NASA space probe **Galileo** took the first detailed photograph of an asteroid while en route to Jupiter. The asteroid photographed, named **Gaspra**, is an irregularly shaped object measuring about 19 by 12 kilometres. The Galileo probe also passed by another asteroid, **Ida**, in August 1993. Both Gaspra and Ida are classified as S-type asteroids because they are composed of metal-rich silicates.

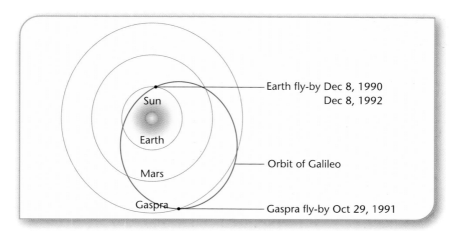

Figure 8.4   Path of the Galileo space probe as it passed by the asteroid Gaspra.

Figure 8.5   The asteroid Gaspra as seen by the Galileo space probe. (Photo: NASA)

Both asteroids are probably fragments of larger parent bodies that were broken apart by catastrophic collisions. Ida's surface is more heavily cratered than Gaspra's, but Ida is much older. Ida has its own companion, **Dactyl**, which is 1.5 km long and orbits Ida at a distance of 100 km.

Figure 8.6   The asteroid Ida and its companion Dactyl, as seen by the Galileo probe. (Photo: NASA)

In 1996, NASA launched the **Near Earth Asteroid Rendezvous** (NEAR) space probe. This probe flew within 1216 km of the asteroid **Mathilde** in 1997. This encounter gave scientists the first close-up look of a carbon-rich C-type asteroid. This visit was unique because NEAR was not designed for fly-by encounters. The next year, NEAR flew past the asteroid **Eros** at a distance of 3829 km, and it went into orbit around Eros in February 2000. In March 2000, the probe was renamed NEAR–Shoemaker in honour of the American astronomer Eugene Shoemaker, who had died not long before. In February 2001, NEAR-Shoemaker became the first spacecraft to land on an asteroid when it landed on Eros.

In October 1998, NASA launched a probe called **Deep Space 1**. The probe flew within 26 km of the asteroid **Braille** in July 1999.

The Japan Aerospace Exploration Agency launched the probe **Hayabusa** on 9 May 2003. The probe's mission was to land on the surface of the asteroid **Itokawa**, and to collect samples and return them to Earth. The first attempt at landing failed, but the second, in September 2005, was successful. Problems with the probe's engines delayed the return flight and the probe is not expected to return to Earth until June 2011. On its return the probe will release its samples via a re-entry capsule. The capsule should land, using a parachute, near Woomera in South Australia.

In September 2007, NASA launched the **DAWN** spacecraft on a mission to the asteroid belt. Seeking clues about the birth of the solar system, the craft will encounter Vesta in 2011 and Ceres in 2015. Both these asteroids are believed to have evolved more than 4.5 billion years ago, about the same time Mercury, Venus, Earth and Mars were formed. Scientists believe the growth of these asteroids was stunted by Jupiter's gravitational attraction. Images taken by the Hubble Space Telescope show these two asteroids are geologically diverse, but mysteries abound. DAWN will orbit each asteroid, photographing the surface and studying their interior composition, density and magnetism. A gamma-ray and neutron detector on board will enable the chemical composition to be measured. A spectrometer will detect visible and infrared light to identify surface

Table 8.2 Significant space probes to asteroids

Probe	Country of origin	Launched	Comments
Galileo	USA	1989	Photographed the asteroid Gaspra in October 1991, photographed the asteroid Ida in August 1993
NEAR–Shoemaker	USA	1996	Passed by Mathilde in 1997, landed on asteroid Eros in 2001
Deep Space 1	USA	1998	Passed by asteroid Braille in 1999
Hayabusa	Japan	2003	Landed on the asteroid Itokawa, returning to Earth in 2011
DAWN	USA	2007	Mission to study Vesta in 2011 and Ceres in 2015

minerals. The DAWN mission is part of NASA's Discovery program – an initiative for lower cost, highly focused, rapid-development scientific spacecraft. The mission marks the first time that a spacecraft will orbit a main-belt asteroid, enabling a detailed and intensive study to be made of it. Dawn will also do a fly-by of Mars in 2009 – the ion engines will then adjust the trajectory to match the orbit of Vesta.

POSITION AND ORBIT

The asteroids orbit the Sun in a region between Mars and Jupiter between 2.0 and 3.5 AU from the Sun. The largest asteroid, Ceres, orbits at an average distance from the Sun of 413 700 000 km (2.76 AU). Ceres is spherical in shape but is not considered a planet because it has not cleared its orbit of other objects; instead it is classed as a dwarf planet. Its diameter is only one-quarter the diameter of Earth's Moon. Pallas orbits the Sun in a slightly elliptical orbit; it is dimmer than Ceres and has a diameter of only 540 km. Vesta is the brightest asteroid, orbiting at an average distance from the Sun of 2.36 AU.

Asteroids within the main asteroid belt are referred to as belt asteroids. In 1857, the American astronomer Daniel Kirkwood suggested that there would be gaps in the asteroid belt, created by gravitational perturbations of Jupiter. The existence of what became known as Kirkwood gaps was confirmed in 1866. Such gaps were made through repeated alignments of asteroids with Jupiter. For example, an asteroid with an orbital period of exactly half that of Jupiter would, on every second orbit, be aligned with Jupiter.

The Trojan asteroids are two groups of asteroids that travel around the Sun in the same orbit as Jupiter. These two groups are held in orbit by the combined gravitational forces of Jupiter and the Sun. About 24 Trojan asteroids have been catalogued so far, but some astronomers believe there may be many more.

The Amor asteroids and Apollo asteroids are two groups that have elongated orbits that take them across the paths of other planets. Apollo asteroids often pass close to Earth. In October 1937, the asteroid **Hermes** came within 900 000 km of our planet. **Eros** is another asteroid that makes regular close approaches to Earth. Such close approaches allow astronomers to examine these asteroids in more detail. As asteroids rotate, different features are visible from Earth, and this enables astronomers to determine their rotational period. Eros, for example, spins once every 5.27 hours.

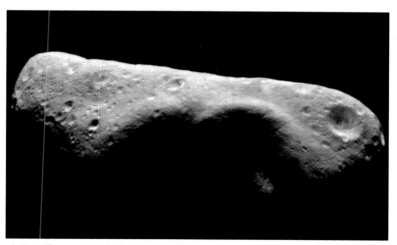

Figure 8.7   The asteroid Eros is irregular in shape because it has experienced a series of collisions. It is 31 km long and its surface is covered by dust and rock fragments. (Photo: NASA)

Although there may be a million asteroids in the asteroid belt, their average separation is actually 10 million kilometres, so collisions are not as common as one might expect.

Some larger asteroids have been found to have smaller asteroid fragments orbiting them. The asteroid **Toutatis** consists of two bodies of similar size orbiting each other.

During 2007, astronomers announced that they have found strong evidence that sunlight can cause asteroids to spin more quickly by 'pushing' on the irregular surface features, and this accelerates or decelerates the rotation rate. The theory is that the Sun's heat serves as a propulsion engine on the irregular features of an asteroid's surface.

DID YOU KNOW?

On 14 June 1968, the asteroid **Icarus** passed within 6 million kilometres of Earth. On 23 March 1989, the asteroid 1989 FC passed within 800 000 km of Earth, and on 9 December 1994, asteroid 1994 XM1 passed within 105 000 km. This latter asteroid is only about 10 m in diameter but it would have done a lot of damage had it hit Earth.

Some asteroids have already hit the Earth's surface. There is a one-kilometre-wide crater in northern Arizona that was made by an asteroid impact some 50 000 years ago. This asteroid was made of iron and nickel and was about 30 m wide. As recently as 1908, what is thought to have been a 100 000-tonne asteroid exploded in the Earth's atmosphere at an altitude of about 10 km with the force of a 20 to 30 megaton nuclear bomb. The explosion occurred over Siberia and the blast from the explosion flattened forests and burned an area about 80 km across. Many scientists believe an asteroid impact about 65 million years ago was responsible for the extinction of the dinosaurs on Earth. The asteroid created a huge circular depression called the Chicxulub Crater centred in Mexico's Yucatan Peninsula. The diameter of the crater is about 180 km.

SIZE AND COMPOSITION

Asteroids vary greatly in size. The largest asteroid, Ceres, is 950 km in diameter and contains about one-third the total mass of all the asteroids. When it was discovered in 1801, Ceres was thought to be another planet because of its size. Ceres was classified by the IAU in 2006 as a 'dwarf planet' because it is too small to have cleared out the smaller chunks of matter orbiting around it in the asteroid belt.

Figure 8.8   The Hubble Space Telescope captured this image of Ceres in 2004. Notice how Ceres is more spherical in shape than the other asteroids. In 2006, Ceres was classified as a dwarf planet because it is part of a larger population of bodies with the same general orbital path. (Photo: NASA)

Figure 8.9   Comparative sizes of the Earth, Moon and Ceres.

The second largest asteroid is Vesta. Vesta has not been classed as a dwarf planet because its shape is not spherical: it has a large concavity and protrusion near its south pole. Vesta is thought to have an iron-nickel core, a rocky olivine mantle and a surface crust containing several impact craters. Fragments of Vesta are thought to be scattered thoughout the solar system.

Scientists have found out about the composition of asteroids by analysing the light reflected from their surface and from analysing meteorite fragments. There are two main types of asteroids, based on their composition. One group dominates the outer part of the belt. These asteroids are rich in carbon and their composition has not changed much since the solar system formed. The second group is located in the inner part of the belt, and is rich in minerals such as iron and nickel. These asteroids formed from melted materials.

Scientists believe Ceres and Vesta have followed different evolutionary paths. Vesta's origins seemed to have been hot and violent because it has basaltic flows on its surface. Ceres seems to have a primitive surface and evidence of water in its minerals. Vesta's physical characteristics reflect those of the inner planets, whereas Ceres is representative of the icy moons of the outer planets. By studying these contrasts and comparing these two asteroids, scientists hope to develop a better understanding of the transition from the rocky inner regions of the solar system to the icy outer regions.

Table 8.3 The 10 largest asteroids

Number	Name	Diameter (km)	Discovered	Discoverer
1	Ceres	950	1801	G. Piazzi
4	Vesta	550	1807	H. Olbers
2	Pallas	540	1802	H. Olbers
10	Hygeia	443	1849	De Gasparis
704	Interamnia	338	1910	V. Cerulli
511	Davida	335	1903	R. Dugan
65	Cybele	311	–	–
52	Europa	291	1858	Goldschmidt
451	Patientia	281	–	–
31	Euphrosyne	270	–	–

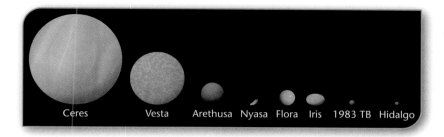

Figure 8.10 Comparative sizes of some asteroids.

THE SURFACE

Scientists have classified asteroids according to the amount of light they reflect. The dark stony kind reflect less than 5 per cent of the sunlight that falls on them. The brighter, light-coloured kind reflect about 20 per cent of incident light. Rarer third kinds resemble iron meteorites and may be the shattered cores of older asteroids.

Photographs of most asteroids show they are covered with craters and dust, made by impact with other smaller rocky bodies. The surfaces of asteroids are generally rock-like, with a layer of soil-like material.

On 12 February 2001, the car-sized NEAR probe landed on the asteroid Eros. Eros is irregular in shape, about 33 km long by 13 km wide. The first images of Eros showed it has an ancient surface covered with craters, grooves, house-sized boulders, and other complex features. The spacecraft's solar panels were able to power NEAR's gamma-ray spectrometer. This instrument analysed material in the surface to a depth of about 10 cm, detecting the elements iron, potassium, silicon and oxygen. The density of Eros is about 2.4 g/cm^3, about the same density as the Earth's crust. Now turned off, the NEAR probe could remain preserved in its present location, the vicinity of the huge, saddle-shaped feature called Himeros, for millions of years. As the asteroid orbits the Sun, the spacecraft's solar panels will be repeatedly turned toward the Sun, offering the possibility of reawakening the probe.

171

DID YOU KNOW?

Observations of Ceres, the largest known asteroid, by NASA's Hubble Space Telescope have revealed that the object may contain pure water beneath its surface. Scientists estimate that if Ceres were composed of 25 per cent water, it may have more water than all the fresh water on Earth. However, unlike the water on Earth, water on Ceres would be in the form of water ice and be located in the mantle. This asteroid is almost round, suggesting it may have an interior with a rocky inner core and a thin, dusty outer crust.

Recent ultraviolet observations by spacecraft have revealed the existence of hydroxide water vapour near the north pole of Ceres. Images of Ceres taken by the Hubble Space Telescope in 2003 and 2004 showed eleven recognisable surface features, the nature of which is currently unknown.

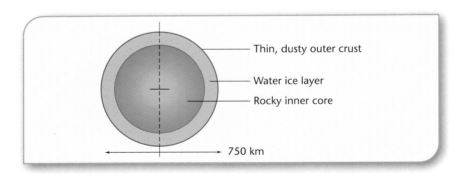

Figure 8.11   Internal structure of Ceres.

THE ATMOSPHERE

With one exception, the asteroids are much too small to retain an atmosphere. Any gases would be lost to space because the very low gravitational attraction is too weak to retain them. Ceres may have a tenuous atmosphere containing water vapour.

TEMPERATURE

The maximum surface temperature of Ceres is −38°C. This is relatively warm compared to the average surface temperature of a typical asteroid of −100°C. This low temperature is largely because of the large distance between the Sun and the asteroid belt. The temperature on Vesta varies from −188°C (dark side) to −18°C (sunlit side). There are no seasons on the asteroids, although they do undergo day and night, dependent on which side is facing the Sun.

MAGNETIC FIELD

The NEAR probe that landed on the asteroid Eros contained a magnetometer to measure any magnetic field. This instrument found no surface magnetic field exists.

It is unlikely that any asteroid would have a magnetic field large enough to be detected.

WEB NOTES

For fact sheets on any of the planets including the asteroids check out:
<http://nssdc.gsfc.nasa.gov/planetary/planetfact.html>
<http://www.nasm.si.edu/etp/>
<http://www.space.com/asteroids/>

CHAPTER 9

JUPITER –
THE GAS GIANT

THE PLANETS MERCURY, Venus, Earth and Mars are regarded as the inner planets of the solar system because they orbit close to the Sun. In contrast, the orbits of the four large planets – Jupiter, Saturn, Uranus and Neptune – are widely spaced at great distances from the Sun. The four inner planets are composed mainly of rock and metal, with surface features such as mountains, craters, canyons and volcanoes. The outer planets, on the other hand, rotate much faster and consist of vast, swirling gas clouds.

The gas planets do not have solid surfaces; their gaseous material simply gets denser with depth. The diameter of such planets is given for levels corresponding to a pressure of one atmosphere. What we see when looking at these planets is the tops of clouds high in their atmosphere.

Jupiter is the first of the gas giants, and the fifth planet from the Sun. This planet is the largest in the solar system. It travels around the Sun once every 11.86 years, at an average distance of 780 million kilometres. Jupiter is so large that over 1300 Earths could be packed into its volume. It is also twice as massive as all the other planets combined (318 times the mass of Earth). Jupiter also contains about 71 per cent of all the material in the solar system, excluding the Sun.

Figure 9.1   The planet Jupiter as seen by the Voyager 1 space probe from a distance of 37 million kilometres. (Photo: NASA)

Jupiter formed from the same swirling mass of gas and dust as the Sun and other planets. But, unlike the inner planets, Jupiter was far enough away from the Sun to retain its envelope of lighter gases, mainly hydrogen and helium. The outer layer of Jupiter forms a gaseous shell almost 20 000 km thick.

Astronomers have studied Jupiter for many years because it is the fourth brightest object in the sky (after the Sun, the Moon and Venus). Ancient observers knew the planet, and its movement across the night sky has been accurately plotted against the background of stars for centuries. Because Jupiter takes about 12 years to orbit the Sun, it spends about a year in each constellation of the zodiac as seen from Earth. To the unaided eye, Jupiter appears as a brilliant white star-like object in the night sky.

Early views about Jupiter

To the ancient Greeks, Jupiter (Greek Zeus) was the king of the gods, the ruler of Olympus and the patron of the Roman state. The planet was also associated with Marduk, the most important figure in Mesopotamian cosmology and the patron god of the city-state of Babylon. According to the story, Marduk fought with Tiamat, the goddess of chaos, and her 11 monsters. Marduk defeated them one by one and split Tiamat's body in two, thus dividing heaven from Earth. Marduk came to symbolise the rule of heavenly order over the universe. The wandering star Jupiter was placed in charge of the night sky.

With the invention of the telescope in 1608, the planet Jupiter could be studied in more detail. In 1610, Galileo observed Jupiter through his telescope and discovered its four largest moons, Io, Europa, Ganymede and Callisto. They are named after some of the mythical lovers and companions of the Greek god Zeus. These moons became known as the **Galilean moons** after Galileo.

The motion of these moons around Jupiter provided evidence to support Copernicus's heliocentric theory of the motions of the planets. Galileo was arrested because of his support for the Copernican theory and he was under house arrest for the rest of his life.

In 1665, the astronomer Giovanni Cassini was the first person to see the Great Red Spot on Jupiter. Twenty-five years later he observed that the speeds of Jupiter's clouds vary with latitude. Nearer the poles, the rotation period of Jupiter's atmosphere is more than five minutes longer than at the equator.

Table 9.1 Details of Jupiter

Distance from Sun	778 330 000 km (5.20 AU)
Diameter	142 984 km
Mass	1.90×10^{27} kg (318 times Earth's mass)
Density	1.33 g/cm^3 or 1330 kg/m^3
Orbital eccentricity	0.048
Period of revolution (length of year)	4329 Earth days or 11.86 Earth years
Rotation period	9 hours 50 minutes
Orbital velocity	47 016 km/h
Tilt of axis	3.12°
Average temperature	−153°C
Number of moons	at least 63
Atmosphere	hydrogen, helium
Strength of gravity	24.6 N/kg at surface

PROBING JUPITER

The first space probe to visit Jupiter was **Pioneer 10**, in December 1973. This was the first probe to venture beyond Mars. The probe returned 23 low-resolution images of Jupiter's cloud system. A year later, **Pioneer 11** returned 17 images during its closest approach to Jupiter; it then used Jupiter's strong gravitational field to propel it towards Saturn. These two probes also recorded data of Jupiter's atmospheric temperature and pressure, and took several pictures of its moons.

The probes recorded changes in Jupiter's atmosphere, particularly around the Great Red Spot, and discovered Jupiter's huge magnetic field.

In 1977, NASA launched the **Voyager 1** and **2** spacecraft on a mission to explore the outer planets. They flew by Jupiter in March and July 1979 before proceeding on to Saturn. After encountering Saturn in November 1980, Voyager 1 headed out of the solar system, while Voyager 2 went on to Uranus and Neptune. Each Voyager was equipped with two television cameras and programmable computers. Voyagers 1 and 2 discovered that Jupiter has complicated atmospheric dynamics, lightning and auroras. These probes also found increased turbulence around the Great Red Spot. The winds to the north and south of the Spot blow in opposite directions,

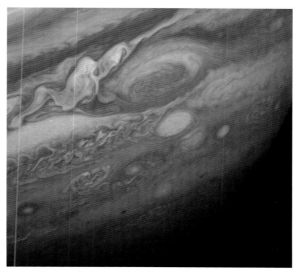

Figure 9.2   View of Jupiter's Great Red Spot taken by Voyager 1 in February 1979. (Photo: NASA)

seemingly fuelling the Spot's rotation. Three new moons were discovered, as well as a ring system.

A surprising finding made by Voyager was that Io, one of Jupiter's moons, has active sulfurous volcanoes on its surface. The surface of Io was found to be very young, with no impact craters being detected. Lava flows were seen extending from some volcanoes. In contrast to Io, the moon Europa was found to be covered by a layer of water ice about 100 km thick. The moons Ganymede and Callisto contained crusts of water ice and large numbers of impact craters.

In 1989, NASA launched the **Galileo** space probe via a space shuttle in Earth orbit. The probe was to rendezvous with Jupiter in 1995, after a trip that used a gravity assist from Venus. In July 1994, while still 225 million kilometres from Jupiter, Galileo was able to observe fragments of the comet Shoemaker-Levy as they hit Jupiter. The fragments hit Jupiter at a speed of about 200 000 km/h over a period lasting about a week. Some of the impacts created plumes or fireballs that were up to 4000 km wide and reached heights of up to 2100 km. The collisions left visible marks in

Jupiter's atmosphere that persisted for months. The markings were also seen from Earth orbit by the Hubble Space Telescope.

When Galileo reached Jupiter in 1995 it released a probe that descended into Jupiter's atmosphere to about 150 km below the cloud tops. Data from this probe indicated that there was much less water than expected. Also surprising was the high temperature and density of the upper parts of the atmosphere. Recent observations by the Galileo orbiter suggest the probe may have entered the atmosphere at one of the warmest and least cloudy areas on Jupiter at that time.

As the **Ulysses** space probe passed by Jupiter in February 1992 it gathered data that showed that the solar wind has a much greater effect on Jupiter's magnetic field than earlier measurements had suggested.

In February 2007, the **New Horizons** probe flew by Jupiter. The reason for the fly-by was to give it a gravitational boost, throwing it towards Pluto. This brief encounter was also used as a test run for both the spacecraft and its earthbound controllers in preparation for the Pluto encounter in 2015. The probe passed within 51 000 km of Jupiter. Images were taken of Jupiter's rings, its moon Io, and the Little Red Spot on Jupiter's surface. New Horizons is the fastest spacecraft ever, having bridged the gap between Earth and Jupiter in only 13 months. Its current velocity is approximately 80 000 km/h.

Table 9.2 Significant space probes to Jupiter

Probe	Country of origin	Launched	Comments
Pioneer 10	USA	1972	First space probe to fly by Jupiter, in 1973
Pioneer 11	USA	1973	Fly-by of Jupiter in 1974, went on to Saturn
Voyager 1	USA	1977	Fly-by of Jupiter in 1979, returned valuable data
Voyager 2	USA	1977	Fly-by of Jupiter in 1979, returned valuable data
Galileo	USA	1989	In orbit around Jupiter 1995; probe dropped into atmosphere
Ulysses	USA, ESA	1990	Fly-by of Jupiter in 1992 while en route to the Sun
New Horizons	USA	2006	Fly-by of Jupiter in 2007 while en route to Pluto

POSITION AND ORBIT

The slightly elliptical orbit of Jupiter lies between the asteroid belt and Saturn. Jupiter has a mean distance from the Sun of just over 778 million kilometres, placing it about 5.2 times further from the Sun than is Earth. It travels around the Sun once every 11.86 years, and it rotates on its axis with a period shorter than any other planet. The short rotational period has resulted in Jupiter becoming flattened or oblate. The equatorial diameter is 142 984 km, which is about 8000 km greater than its polar diameter. This shape suggests the interior is a liquid rather than a solid or gas. The planet's axis that is tilted at an angle of 3.12° to the vertical.

DENSITY AND COMPOSITION

Jupiter is more than 318 times more massive than the Earth and has twice as much mass as all the other planets combined. Jupiter emits about twice

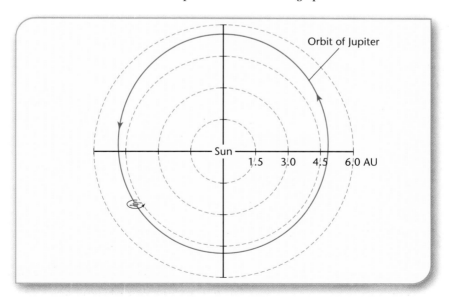

Figure 9.3   Orbital path of Jupiter (distance circles are in astronomical units, AU).

as much heat as it absorbs from the Sun. Its core temperature is estimated at 20 000°C – about three times greater than Earth's core temperature. This heat is thought to be generated by the gravitational contraction of Jupiter by about 3 cm per year. The core pressure may be about 100 million times greater than on Earth's surface. Although its interior temperatures and pressures are very high, it does not have enough mass to become a star.

Jupiter's average density is lower than that of Earth – only 1.33 g/cm³ compared to Earth's 5.52 g/cm³. The shape of the planet and its strong gravitational field suggest Jupiter must have a dense core about 10 to 20 times the mass of Earth.

Jupiter is about 90 per cent hydrogen and 10 per cent helium, with traces of methane, water, ammonia and 'rock'. This composition is very close to what is thought to have been the composition of the nebula from which the solar system formed.

There are three main regions to Jupiter's interior. The outer layer of Jupiter that we see from Earth is the top of the outer layer of clouds. The Galileo probe found that these clouds are mostly gaseous molecular hydrogen and helium. With increasing depth and hence pressure, the gases become more like liquids. The hydrogen becomes crushed into a

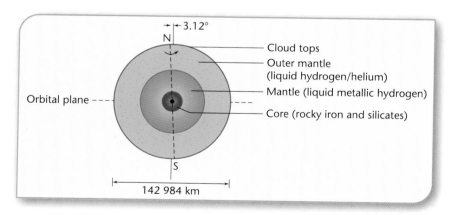

Figure 9.4 Interior structure of Jupiter.

liquid form called metallic hydrogen. Metallic hydrogen's high electrical conductivity, combined with the rapid rotation of the planet, give rise to Jupiter's fierce magnetic field and radiation belts. Most of the interior of Jupiter (its mantle) is therefore mostly liquid metallic hydrogen. Below the mantle, the third region or core of Jupiter is thought to consist of rocky material, mainly iron and silicates. The core is about 20 000 km in diameter.

The strength of gravity on Jupiter is 2.5 times Earth's gravity. This means that a 75 kg person who weighs 735 N on Earth would weigh 1845 N on Jupiter.

The surface

The outer layers of Jupiter form a shell, mostly of gaseous hydrogen. Although this layer is about 20 000 km thick, there is no solid surface. The gas just gets thicker and thicker, until the pressure is 3 million times the air pressure at sea level on Earth. At this point, hydrogen becomes crushed into liquid metallic hydrogen.

Seen from Earth, Jupiter is one of the brightest planets in the sky. Viewed through a telescope, its disc is seen to be crossed by numerous belts or

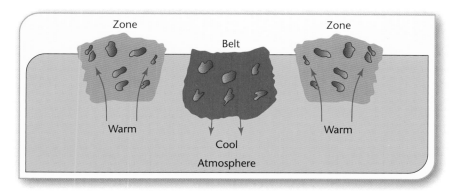

Figure 9.5   Convection currents and the planet's rapid rotation create the bright zones and dark belts surrounding Jupiter.

zones of various colours, including red, orange, brown and yellow. The brighter zones are regions where fluids from within the planet are rising to the surface to cool, while the darker belts are regions where material is descending.

Also visible on Jupiter is the Great Red Spot, first observed by Giovanni Cassini through his telescope in 1665. This Spot is a huge, oval-shaped atmospheric feature located in the southern hemisphere. The size of the Spot varies, but it is roughly 30 000 km long and 12 000 km wide. The Pioneer and Voyager missions suggested that the Spot is a hurricane-like storm whose red colour may be caused by the presence of red phosphorus and yellow sulfur in ammonia crystals. Infrared observations show the Spot is a high-pressure region whose cloud tops are much higher and colder than the surrounding regions. It is not fully understood how the Spot lasts for so long, as it must be absorbing a lot of energy to keep surviving. The

DID YOU KNOW?

In March 1993, astronomers Eugene Shoemaker, Carolyn Shoemaker and David Levy discovered a comet near Jupiter. When the comet was discovered, it had broken into 21 pieces, probably stretched by Jupiter's gravity until it shattered. In July 1994, the fragments of the comet, named Shoemaker-Levy 9, crashed into Jupiter at an estimated speed of 200 000 km/h. The impacts lasted almost a week and were observed by the Hubble Space Telescope from Earth and by the Galileo space probe, which was on its way to Jupiter. Some of the impacts created plumes or fireballs that were up to 4000 km wide and reached heights of up to 2100 km. These plumes flattened out and fell back on to the cloud tops of Jupiter's atmosphere. The largest collision scattered material over an area almost as large as the Earth. The collisions were clearly visible as dark spots for months after the impacts. The darkness may have come from carbon compounds from the comet.

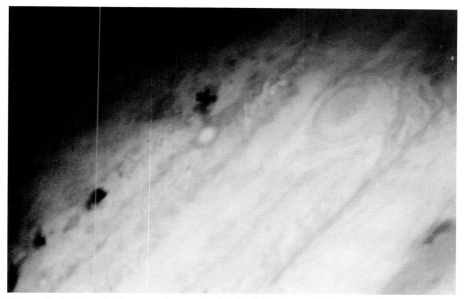

Figure 9.6   The dark spots shown in this Hubble Space Telescope photograph are the impact scars of fragments of comet Shoemaker-Levy 9 as they hit Jupiter in July 1994. (Photo: NASA)

edge of the Spot rotates at a speed of about 360 km/h. The Spot moves east to west around Jupiter but stays about the same distance from the equator.

THE ATMOSPHERE

Spectral analysis of Jupiter's atmosphere shows it is 86 per cent hydrogen by mass and 13 per cent helium. The remainder consists of small amounts of simple compounds such as methane, ammonia, and water vapour.

Data from the Galileo probe released into Jupiter's atmosphere goes down to only about 150 km below the cloud tops. It is thought that the high pressure and radiation on Jupiter destroyed the Galileo probe's sensors. Three distinct layers of clouds are thought to exist on Jupiter. The upper layer is the coldest (−153°C) and contains mainly ammonia ice crystals. The middle layer contains crystals of ammonium hydrosulfide and a mixture of ammonia and hydrogen sulfide. The lower layer contains water ice. The vivid colours seen in Jupiter's clouds probably result from chemical reactions between the compounds in the atmosphere. The colours seem to correlate with altitude: blue for the lowest clouds, followed by browns and whites, with reds highest.

Clouds in Jupiter's turbulent atmosphere move at high velocity in east–west belts parallel to the equator. The winds blow in opposite directions in adjacent belts. Data from the Galileo probe indicate the winds travel at about 600 km/h and extend down thousands of kilometres into the interior. The winds are mainly driven by Jupiter's internal heat and the planet's rapid rotation.

The Pioneer and Voyager probes sent back beautiful images of Jupiter's atmosphere. Comparison of pictures from the two probes showed changes

Figure 9.7   Close-up view of Jupiter's clouds taken by Voyager (red = high cloud, brown, white = middle-level cloud, blue = low cloud). (Photo: NASA)

are occurring in Jupiter's atmosphere. The most notable of these was in the region around the Great Red Spot. At the time of the Pioneer probes (1973–74), the Spot was surrounded by a white zone. By 1979, when Voyager visited Jupiter, a dark belt had crossed the Spot, and there was increased turbulence around the area. Measurements showed the Spot rotates anti-clockwise over a period of about six days. The winds north of the Spot blow in the opposite direction to those south of the Spot. Lightning 10 000 times more powerful than any seen on Earth has also been detected in Jupiter's atmosphere.

Other oval-shaped features called eddies or circular winds can be seen in the atmosphere of Jupiter. These eddies move about within the zones in which they are trapped by opposing winds. They usually appear white in colour. The Great Red Spot is a huge eddy.

DID YOU KNOW?

Jupiter has a system of three rings surrounding it in an equatorial plane. These rings are much fainter and lighter than those around Saturn and can't be seen from Earth through normal telescopes. The rings were discovered by the Voyager probes, but they have since been imaged in the infrared from ground-based telescopes and by the Galileo probe.

The main ring is about 7000 km wide and 30 km thick. Its inner edge is about 123 000 km above Jupiter's cloud tops.

The rings are dark and are composed of very fine-grained dust particles and rock fragments and, unlike Saturn's rings, they do not contain any ice. Galileo found evidence that the particles are continuously being kicked out of orbit by radiation from Jupiter and the Sun. The rings are probably re-supplied with material by dust particles formed by micrometeor impacts on the four inner moons.

TEMPERATURE AND SEASONS

The temperature near the cloud tops of Jupiter measures about −153°C. Temperatures increase with depth below the clouds, reaching 20°C at a level where the atmospheric pressure is about 10 times as great as it is on Earth. At a depth of about 20 000 km below the cloud tops the temperature is about 10 000°C. Below this depth, the pressure and temperature are high enough to transform liquid hydrogen into liquid metallic hydrogen. The pressure at Jupiter's centre is about 80 million atmospheres and the temperature about 24 000°C, which is hotter than the surface of the Sun.

Because Jupiter takes 11.86 years to orbit the Sun and it has a small axial tilt, there are not any real seasons on Jupiter.

MAGNETIC FIELD

Jupiter has a huge magnetic field, about 20 times stronger than Earth's magnetic field. Its magnetosphere extends only a few million kilometres towards the Sun, but away from the Sun it extends to a distance of more

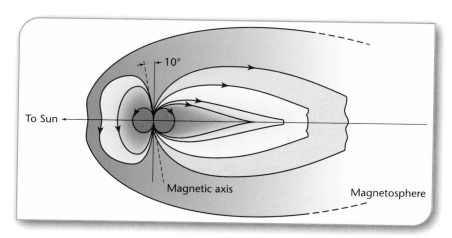

Figure 9.8   Jupiter's magnetosphere is pushed by the solar wind. The magnetic poles are offset from the rotational axis by nearly 10°.

than 650 million kilometres (past the orbit of Saturn). Many of the moons of Jupiter lie within its magnetosphere. The magnetic field contains high levels of radiation and energetic particles, so entering the region it covers would be fatal to space travellers. The Galileo probe discovered a new intense radiation belt between Jupiter's rings and the uppermost atmospheric layers. This belt is about 10 times more intense than Earth's Van Allen radiation belts, and contains high-energy helium ions.

Jupiter emits radio waves strong enough to be picked up by radiotelescopes on Earth. Scientists have used these waves to calculate the rotational speed of Jupiter. The strength of the waves varies as the planet rotates and is influenced by Jupiter's magnetic field. The radio waves come in two forms. The first of these is a continuous emission from Jupiter's surface; the second is a strong burst that occurs when the moon Io passes through certain regions of the magnetic field and radiation belt.

X-ray telescopes and the Hubble Space Telescope regularly detect auroras on Jupiter. These auroras are thousands of times more powerful than those on Earth. On Earth, the most intense auroras are caused by outbursts of charged particles from the Sun interacting with the polar magnetic field of Earth. On Jupiter, however, the particles seem to come from the moon Io, which has volcanoes that spew out oxygen and sulfur ions. Jupiter's strong magnetic field produces about 10 million volts around its poles, and this field captures the charged particles and slams them into the planet's atmosphere. The particles interact with molecules in the atmosphere and the result is intense X-ray auroras, virtually all the time.

MOONS OF JUPITER

At least 63 natural satellites or moons orbit Jupiter. Galileo was the first to observe the four largest moons of Jupiter when he examined the planet through his telescope in 1610. These moons, now known as the Galilean moons, orbit Jupiter in an equatorial plane, creating the appearance of a

Table 9.3 Details of the Galilean moons of Jupiter

Name	Distance from Jupiter (km)	Period (days)	Diameter (km)	Discovered
Io	421 700	1.77	3630	1610
Europa	671 000	3.55	3138	1610
Ganymede	1 070 000	7.16	5262	1610
Callisto	1 883 000	16.69	4800	1610

mini solar system. In order of distance from Jupiter, the four Galilean moons are Io, Europa, Ganymede, and Callisto.

The first three Galilean moons are locked together into a 1 : 2 : 4 orbital resonance by tidal forces from Jupiter. In a few hundred million years Callisto will be locked in too, orbiting at exactly twice the period of Ganymede and eight times the period of Io.

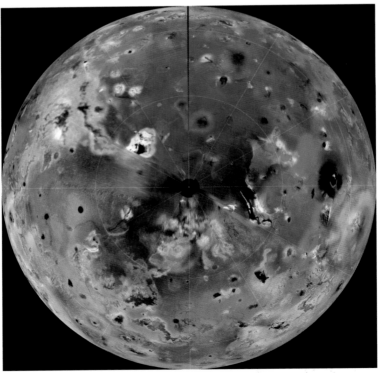

Figure 9.9 Io is the most volcanically active moon in the solar system. (Photo: NASA)

Io is about the same size as Earth's Moon and it orbits Jupiter once every 1.77 days. This moon is the most volcanically active body in the solar system. Images from Voyager 1 showed Io has nine giant erupting volcanoes on its surface and up to 200 smaller volcanoes. These sulfurous eruptions give Io a white-, yellow- and orange-coloured surface.

Io's volcanoes are all relatively flat. The largest volcano, called Pele, is about 1400 km across. There are also mountains up to 10 km high, but these are not volcanic. Voyager also discovered numerous black spots scattered across Io; these are thought to be volcanic vents through which eruptions occur. The volcanic eruptions change rapidly. In the four months between the arrivals of Voyager 1 and Voyager 2, some of the eruptions had stopped while others had begun, and deposits around the vents also changed visibly.

Io is thought to be volcanically active because of the huge tidal forces created by Jupiter. Jupiter's gravity pulls the surface of Io so much that the surface flexes, or bends back and forth. This movement generates enough heat to melt the interior and produce Io's hot-spot volcanism. Io's thin atmosphere is mostly sulfur dioxide gas produced by the volcanoes. At night, some of this gas freezes, and produces the white areas seen on the surface. Io has little or no water.

Europa is the second of the Galilean moons and the fourth largest moon of Jupiter. It is slightly smaller than Earth's Moon and orbits Jupiter once

Figure 9.10   Io as seen by Voyager 1, showing an enormous volcanic eruption in the upper left corner. This material was thrown to a height of 100 km. (Photo: NASA)

every 3.55 days. Io and Europa have a similar composition, consisting of mainly silicate rock. Unlike Io, Europa has a thin outer layer of ice and a layered internal structure, probably with a metallic core. Its surface is relatively smooth, with no mountains and very few craters. Some astronomers think topographical features and a layer of liquid water may exist below the ice-covered surface. Streaks and cracks in the surface may be caused by tidal forces generated by Jupiter.

In 1995, astronomers discovered the atmosphere of Europa is very thin and contains molecular oxygen. This oxygen is thought to be generated by sunlight and charged particles hitting Europa's icy surface, producing water vapour that is split into hydrogen and oxygen. The hydrogen escapes, leaving behind the oxygen.

The Galileo probe found that Europa has a weak magnetic field.

Ganymede is the third of the Galilean moons and is the largest moon in the solar system. Its diameter is larger than the planet Mercury, although its mass and density is much less. Ganymede orbits Jupiter in

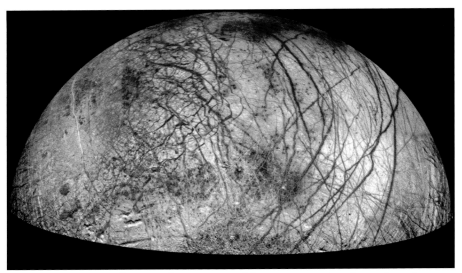

Figure 9.11　Europa's icy surface is covered by numerous streaks and cracks (photo taken by Galileo space probe). (Photo: NASA)

synchronous rotation once every 7.16 days at a distance of about one million kilometres.

Ganymede has both dark and light areas on its surface. The dark areas are old and heavily cratered. The lighter regions are young and have few impact craters, but they have many grooves and ridges. The largest feature on Ganymede is Galileo Regio, a dark circular area of ancient crust 4000 km in diameter that contains an abundance of craters.

The crust is thought to be about 75 km thick and contain an outer layer of ice. Beneath lies a mantle of either water or ice, and a rocky or silicate-rich core. The Hubble Space Telescope has recently found evidence of oxygen in Ganymede's atmosphere, very similar to Europa's atmosphere.

The Galileo space probe found that Ganymede has its own magnetic field embedded inside Jupiter's huge field. This is probably generated in a similar fashion to the Earth's, as a result of moving, conducting material in the interior.

Callisto is the fourth of the Galilean moons and the second largest moon orbiting Jupiter. It is only slightly smaller than the planet Mercury, but has

Figure 9.12   Ganymede's surface has some smooth areas and some cratered and furrowed areas. (Photo: NASA)

only one-third its mass. Callisto orbits Jupiter once every 16.69 days in synchronous rotation at a distance of about two million kilometres.

Callisto's surface is a dark, ancient and icy crust, covered with many old impact craters. It is the most heavily cratered object in the solar system. The craters and impact basins are relatively flat because of the nature of the surface. The largest impact basin, Valhalla, is about 3000 km in diameter and is surrounded by bright concentric rings of fractured ice. Valhalla may have formed as a result of a huge asteroid impact. Numerous smaller craters cover this feature, which suggests an age of about four billion years.

Callisto's craters lack the high ring mountains, radial rays and central depressions common in craters on Earth's Moon and Mercury. Detailed images from the Galileo probe show that the craters have been obliterated in some areas.

Unlike Ganymede, Callisto has no grooved terrain, suggesting little if any tectonic activity has occurred. It probably cooled very rapidly. Voyager's instruments measured a temperature range from −180°C during daytime to −315°C at night.

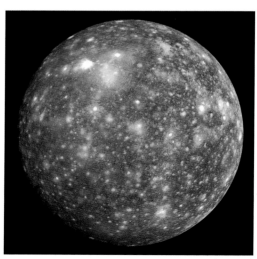

Figure 9.13   Callisto is the most heavily cratered moon in the solar system. It is geologically dead.

Data from the Galileo probe suggests Callisto has little internal structure, and is composed of about 40 per cent ice and 60 per cent rocky iron. Its atmosphere is very tenuous and contains mainly carbon dioxide. There is no evidence of a magnetic field.

OTHER SATELLITES OF JUPITER

In addition to the four Galilean moons, Jupiter has at least 59 other moons. The non-Galilean moons tend to be irregular in shape and smaller than 185 km in diameter. Four of these moons are closer to Jupiter than Io, the rest orbit in regions beyond Callisto. The outer bodies are probably captured asteroids, while the inner ones are probably pieces broken off a larger body.

The four inner moons are **Metis**, **Adrastea**, **Amalthea** and **Thebe**. The largest of these is Amalthea, which was discovered in 1892 by the American astronomer Edward Barnard, and is only 167 km across. Amalthea was originally thought to be innermost moon, but the Voyager probes found Metis and Adrastea to be closer to Jupiter. Amalthea is a dark, heavily cratered, irregularly shaped body with a reddish colour. Metis, Adrastea and Thebe were discovered by the Voyager 1 probe in 1979.

The four inner moons and the Galilean moons all orbit in an equatorial plane in near-circular orbits. The outer moons have more elliptical orbits and are more inclined to Jupiter's orbital plane. All the moons after Carpo orbit in the opposite direction to that of the other satellites, and opposite

Table 9.4 Details of the inner moons of Jupiter

Name	Distance from Jupiter (km)	Period (days)	Diameter (km)	Discovered
Metis	127 690	0.29	43	1979
Adrastea	128 690	0.30	16	1979
Amalthea	181 400	0.50	167	1892
Thebe	221 900	0.68	98	1979

to the direction of Jupiter's spin. These characteristics, together with their small size, suggest that these outer satellites are captured asteroids, not originally part of the Jupiter system.

DID YOU KNOW?

Between 9 and 11 December 2001, 10 moons of Jupiter were discovered, bringing the total to 39. During the following year (2002) only one new moon was discovered. However, four months later, between 5 and 9 February 2003, 23 more moons were found, bringing the total to 63 moons.

Table 9.5 Details of the outer moons of Jupiter

Name	Distance from Jupiter (km)[a]	Period (days)[b]	Diameter (km)	Discovered
Themisto	7 284 000	130	8	2000
Leda	11 165 000	241	20	1974
Himalia	11 461 000	250	170	1904
Lysithea	11 717 000	260	36	1938
Elara	11 741 000	261	86	1905
S/2000J11	12 555 000	287	4	2000
Carpo	17 144 000	458	3	2003
S/2003J12	17 740 000	482	1	2003
Euporie	19 090 000	538	2	2001
S/2003J3	19 621 000	561	2	2003
S/2003J18	19 812 000	570	2	2003
Thelxinoe	20 453 000	597	2	2004
Euanthe	20 464 000	598	3	2001
Helike	20 540 000	601	4	2003
Orthosie	20 570 000	602	2	2001
Iocaste	20 722 000	609	5	2000
S/2003J16	20 743 000	610	2	2003
Praxidike	20 823 000	613	7	2000
Harpalyke	21 063 000	624	4	2000
Mneme	21 129 000	627	2	2003

Name	Distance from Jupiter (km)[a]	Period (days)[b]	Diameter (km)	Discovered
Hermippe	21 182 000	630	4	2004
Thyone	20 939 000	640	4	2001
Ananka	21 405 000	642	28	1951
S/2003J17	22 134 000	672	2	2003
Aitne	22 285 000	679	3	2001
Kale	22 409 000	685	2	2001
Taygete	22 438 000	687	5	2000
S/2003J19	22 709 000	699	2	2003
Chaldene	22 713 000	699	4	2000
S/2003J15	22 720 000	700	2	2003
S/2003J10	22 730 000	700	2	2003
S/2003J23	22 739 000	700	2	2003
Erinome	22 986 000	712	3	2000
Aoede	23 044 000	715	4	2003
Kallichore	23 111 000	718	2	2003
Kalyke	23 180 000	721	5	2000
Carme	23 197 000	722	46	1938
Callirrhoe	23 214 000	722	9	1999
Eurydome	23 230 000	723	3	2001
Pasithee	23 307 000	727	2	2001
Cyllene	23 396 000	731	2	2003
Eukelade	23 483 000	735	4	2003
S/2003J4	23 570 000	739	2	2003
Pasiphae	23 609 000	741	60	1908
Hegemome	23 703 000	745	3	2003
Arche	22 717 000	746	3	2002
Isonoe	23 800 000	750	4	2000
S/2003J9	23 857 000	753	1	2003
S/2003J5	23 973 000	758	4	2003
Sinope	24 057 000	762	38	1914
Sponde	23 252 000	772	2	2001
Autonoe	24 264 000	772	4	2001
Kore	23 345 000	776	2	2003
Megaclite	24 687 000	792	5	2000
S/2003J2	30 290 000	1077	2	2003

a   Distances are given as per mean semi-major axis.
b   Periods are given to the nearest day.

Jupiter's outer moons are probably the remnants of a single asteroid that was captured by Jupiter and broken up.

As of 2007, Jupiter has four small inner moons, four large Galilean moons, and 55 other tiny satellites of sizes ranging from 1 km to 170 km across. The tiny 'asteroid-like' satellites orbit between 7 284 000 km and 30 290 000 km distant from the planet, and may be better called 'moonlets'.

Web notes

For fact sheets on any of the planets, including Jupiter, check out:
    <http://nssdc.gsfc.nasa.gov/planetary/planetfact.html>
    <http://www.space.com/jupiter/>
    <http://www.nasm.si.edu/etp/>

CHAPTER 10

SATURN –
THE RINGED PLANET

SATURN IS THE second of the gas giants and the sixth planet from the Sun. This planet is the second largest in the solar system, with a diameter of 120 536 km. It travels around the Sun once every 29.46 years at an average distance of 1430 million kilometres. At its closest approach to the Earth, Saturn is about 1 278 000 km away.

Saturn travels around the Sun in an elliptical orbit. Its distance from the Sun varies from about 1509 million kilometres at its furthest point to about 1350 million kilometres at its closest point.

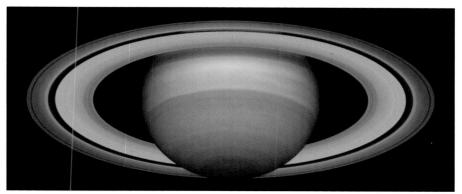

Figure 10.1   Saturn and its ring system as seen by Voyager 2. (Photo: NASA)

Saturn is about 85 per cent the size of Jupiter but is twice as far from Earth. It is over 95 times as massive as the Earth and, with the exception of Jupiter, has more mass than all the other planets combined. However, Saturn has the lowest density of all the planets, only 0.69 g/cm^3, which is less than the density of water and roughly half the density of Jupiter. This low density means Saturn must be composed of light elements.

Saturn formed from the same swirling mass of gas and dust as the Sun and other planets. Like Jupiter, and unlike the inner planets, Saturn was far enough away from the Sun to retain its envelope of lighter gases, mainly hydrogen and helium. As it orbits the Sun, Saturn spins on its axis, faster than any other planet. Its axis is tilted at an angle of 26.73° from the perpendicular.

Astronomers have studied Saturn for many years, as it is easily visible in the night sky of Earth. Ancient observers knew it, and its movement across the night sky has been accurately plotted against the background of stars for centuries. To the unaided eye, Saturn appears as a brilliant yellow-orange star-like object in the night sky. Saturn's main feature is its spectacular ring system, which can be seen through a telescope from Earth. The only other planets to have rings are Jupiter, Neptune and Uranus, but their rings aren't as prominent as Saturn's rings and they can't be seen through a telescope from Earth.

EARLY VIEWS ABOUT SATURN

Saturn has been observed in the night sky since prehistoric times. Mesopotamian astronomers called Saturn the 'the old sheep' or 'the eldest old sheep', while the Assyrians described the planet as a sparkle in the night sky and named it 'Star of Ninib'. To the ancient Romans, Saturn was the god of agriculture, while the Greeks called it Cronos, after Zeus's father, the overthrown ruler of the universe. Cronos was the also son of Uranus and Gaia. Saturn is the root of the English word 'Saturday'.

Table 10.1 Details of Saturn

Distance from Sun	1 430 000 000 km (9.54 AU)
Diameter	120 536 km
Mass	5.68×10^{26} kg (95.18 times Earth's mass)
Density	0.69 g/cm³ or 690 kg/m³
Orbital eccentricity	0.056
Period of revolution (length of year)	10 768 Earth days or 29.46 Earth years
Rotation period	10 hours 40 minutes
Orbital velocity	34 704 km/h
Tilt of axis	26.73°
Average temperature	–185°C
Number of moons	62
Atmosphere	hydrogen, helium
Strength of gravity	10.4 N/kg at surface

The modern era for Saturn began in 1610, when Galileo first observed it through his telescope and described it as a triple-bodied object. Other observers thought Saturn had 'handles' or 'ears'. In 1659, Christiaan Huygens reported that Saturn was circled by a broad, flat ring and had a moon; this was to be called Titan. In 1676, the astronomer Giovanni Cassini discovered a gap in Saturn's ring system. With modern telescopes, Earth-based astronomers have found Saturn has two prominent rings (A and B), and one faint ring (C). The gap between the A and B rings is now known as the **Cassini division**. A much fainter gap in the A ring is known as the **Encke division** after the German astronomer Johann Franz Encke, who allegedly saw it in 1838. Pictures taken by the Voyager probes show four additional faint rings.

PROBING SATURN

People on Earth have observed Saturn through telescopes based on Earth since 1610 when Galileo first observed it, and more recently through space-based telescopes.

The first space probe to visit Saturn was **Pioneer 11**, on 1 September 1979. The probe passed within 21 000 km of the planet and within 3500 km

Figure 10.2   Enhanced-colour photograph taken by Voyager of Saturn's rings. Different colours denote differences in chemical composition and structure of the ring particles. (Photo: NASA)

of its outer ring. It travelled under the ring system and sent back many useful pictures of the rings. However, the images provided little new information about Saturn's clouds and atmosphere.

In November 1980, **Voyager 1** passed within 124 123 km of Saturn before moving out of the solar system. **Voyager 2** encountered Saturn on 26 August 1981, getting to within 101 335 km of the planet before proceeding to Uranus and Neptune. The Voyager probes provided many pictures and a large quantity of data about Saturn. They found three new moons and four additional faint rings around Saturn, and provided pictures of atmospheric circulation. Before the Voyager probes, information about Saturn's atmosphere was limited because astronomers could see only the tops of the clouds from Earth. The Voyager probes identified long-lived, oval-shaped structures inside the clouds, and revealed three layers of clouds with slightly different compositions.

In 1994, the Hubble Space Telescope captured the first images of auroras in Saturn's atmosphere. It also captured images of topographical features on Saturn's largest moon, Titan, which suggest a continent once existed on this moon.

The **Cassini** mission to Saturn (named after Giovanni Cassini) was one of the most ambitious ever attempted. It was a joint venture of NASA, the European Space Agency (ESA) and the Italian Space Agency (known as ASI for its acronym in Italian), and was designed to explore the whole Saturnian system, the planet itself, its atmosphere, rings and magnetosphere, and some of its moons. Launched in 1997, the Cassini space probe reached Saturn in 2004 and went into orbit around the planet. Cassini plunged between Saturn's two outer rings at 80 000 km/h before it slowed down enough to be captured by Saturn's gravity and began orbiting the planet. Instruments on board Cassini detected an eruption of atomic oxygen in Saturn's E ring.

Cassini has also taken pictures of Saturn's largest moon, Titan. In July 2004, Titan was found to be surrounded by a thick atmosphere, with areas of water ice on its surface. Cassini also released a smaller probe called **Huygens** on 24 December 2004. Twenty days later, the probe entered Titan's atmosphere at about 6 km/s, and landed via parachute on Titan's surface, on 14 January 2005. The probe landed in mud-like wet clay covered by a thin crust. The first images showed a pale orange, rock-strewn, eroded landscape with drainage channels. The ground temperature was a chilling −180°C. Huygens was the first successful attempt by humans to land a probe on another world in the outer solar system.

By the end of 2007, Cassini had flown by Titan forty times and mapped over 60 per cent of Titan's 'Lake District', north of latitude 60°. Dark areas on the surface are believed to be filled with a mixture of liquid ethane, methane, and dissolved nitrogen. Some of the lakes appear to be fed by

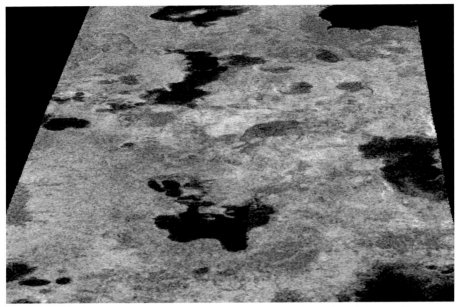

Figure 10.3   Radar image taken by Cassini of the hydrocarbon lakes on Titan, Saturn's largest moon. (Photo: NASA)

rivers that flow down from the surrounding hill country to shorelines of bays, peninsulas and islands. The rivers have many tributaries among the uplands. A few lakebeds appear dry.

POSITION AND ORBIT

Saturn is the second largest planetary member of the solar system and is the sixth planet from the Sun. Its orbit is slightly elliptical and lies between those of Jupiter and Uranus. Saturn has a mean distance from the Sun of

Table 10.2 Significant space probes to Saturn

Probe	Country of origin	Launched	Comments
Pioneer 11	USA	1973	First space probe to fly by Saturn in 1979
Voyager 1	USA	1977	Passed by Saturn in November 1980
Voyager 2	USA	1977	Passed by Saturn in August 1981
Cassini	USA, ESA, ASI	1997	Began orbiting Saturn in 2004

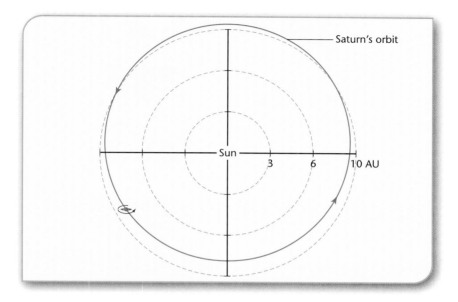

Figure 10.4 Orbital path of Saturn (distance circles are in astronomical units, AU).

just over 1430 million kilometres, placing it about 9.5 times further from the Sun than is Earth. It travels around the Sun once every 29.46 years and rotates on its axis with a period of 10 hours 40 minutes. As with Jupiter, the short rotational period has resulted in Saturn becoming flattened or oblate. The equatorial diameter is 120 536 km, which is 10 per cent more than its polar diameter of 108 728 km. This shape suggests the interior is a liquid rather than a solid or gas.

DENSITY AND COMPOSITION

Saturn is the least dense of all the planets, mainly because of its composition of light gases. Like Jupiter, Saturn is a gaseous planet, composed of about 75 per cent hydrogen and 25 per cent helium, with traces of water, methane, ammonia and 'rock' that is believed to be similar to the composition of the primordial solar nebula from which the solar system was formed. Saturn's mass is about 95 times greater than Earth's but it has 800 times the volume.

The interior of Saturn is also similar to that of Jupiter, in that is contains a rocky core, a liquid metallic hydrogen mantle, and a liquid outer layer of molecular hydrogen. Traces of various ices are also present. With increasing height, the outer layer of liquid hydrogen becomes gaseous.

Because of its smaller mass, Saturn's interior is less compressed than Jupiter's. As a result, Saturn's rocky core is larger than Jupiter's core, and the pressure is unable to convert much hydrogen into a liquid metal. The core is about 32 000 km in diameter, while the mantle is about 12 000 km thick. Saturn's core contains around 26 per cent of the total mass of the planet, as opposed to around 4 per cent for Jupiter.

As with Jupiter, Saturn's interior is hot, about 12 000°C at the core, and the planet radiates more energy into space than it receives from the Sun.

The strength of gravity on Saturn is only slightly more than Earth's gravity (10.4 N/kg compared to 9.8 N/kg). This means that a 75 kg person weighs 735 N on Earth, but on Saturn would weigh 780 N.

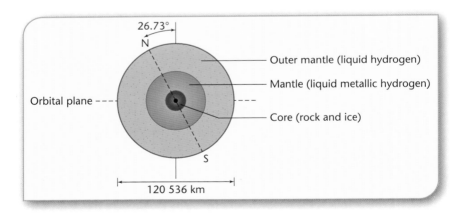

Figure 10.5   Internal structure of Saturn.

THE SURFACE

Saturn's surface and interior are similar to those of Jupiter. Saturn's mantle is surrounded by ordinary liquid hydrogen, so there is no solid surface layer. When observing Saturn we are looking at its cloud tops. These cloud tops lack the colours visible on Jupiter. Photographs taken by Voyager 1 show faint bands across Saturn's surface but these are nowhere near as prominent as those on Jupiter. This is mainly because the gravitational pull of Saturn is much weaker than Jupiter's, and hence the layers of gases are more weakly held together. The banded appearance of the cloud layer is thought to be caused by differences in the temperature and altitude of the atmospheric gas masses.

Saturn does have some long-lived spots like the Great Red Spot on Jupiter. In 1990, the Hubble Space Telescope observed an enormous white cloud near Saturn's equator that was not present during the Voyager encounters. In November 1994, another spot was observed near Saturn's equator. This storm-like spot was 12 700 km across – about the same size as Earth.

THE ATMOSPHERE

Saturn's atmosphere contains about 88 per cent hydrogen, 11 per cent helium, and 1 per cent other elements. The atmosphere consists of three cloud layers, similar to Jupiter's, but the temperature is generally lower.

The lower layer of the clouds contains water ice crystals. The middle layer contains clouds of ammonium hydrosulfide, while the uppermost layer contains ammonia ice crystals. The clouds generally rotate with Saturn, with a period of 10 hours 14 minutes at the equator and 10 hours 40 minutes at high latitudes.

The clouds appear yellow in colour and move in zones parallel to the equator, with winds that alternate from east to west between zones. Wind speeds are generally higher than those on Jupiter. A fast equatorial flow like a giant jet stream has been clocked on Saturn. This high-velocity wind of 1800 km/h has remained fairly constant over decades. Saturn also has storms like those seen on Jupiter, but they are less visible and less frequent, although they last longer.

Because Saturn is 9.53 times further from the Sun than is Earth, its atmosphere receives only 1 per cent of the solar energy Earth receives. It radiates twice this amount of energy from its interior. Gases being heated by the interior and Saturn's fast rotation generate circulation patterns in the atmosphere.

THE RINGS

The most prominent feature of Saturn is its ring system, which encircles the planet around its equator. The rings do not touch Saturn. As Saturn orbits the Sun, the rings tilt at the same angle as the equator. Sometimes we see the rings edge-on from Earth and sometimes they are nearly upright. The Voyager space probes showed much more detail about the rings than could be seen from Earth, and four additional rings were discovered, bringing the total to seven.

Figure 10.6   Infrared photograph by Cassini of shadows cast by Saturn's A, B and C rings on the planet's surface. (Photo: NASA)

Closer examination of the rings by space probes has revealed they are actually composed of hundreds of narrow, closely spaced 'ringlets'.

The closest ring to Saturn is the faint D ring. The inner edge of this ring lies about 6700 km from the cloud tops. The C ring begins at about 14 200 km altitude. The densest of the rings is the B ring, which begins at about 31 700 km above the cloud tops. The Cassini division, which is a gap of about 4700 km in width, is at an altitude of 57 200 km. The gap probably formed as a result of the gravitational pull between ring particles and Saturn's moon Mimas. Beyond this gap, the A ring begins at 76 500 km altitude, followed by the narrow, faint, F ring. Beyond the F ring is the tenuous G ring, discovered by the Voyager space probe, and the even more tenuous E ring (the outermost ring). The bulk of the ring system spans about 275 000 km but it is only about 1.5 km thick.

Saturn's rings, unlike the rings of other planets, are very bright because they reflect light. The rings consist of countless small particles, ranging in size from a centimetre to 10 metres across. In order to measure the size of the particles in Saturn's rings, scientists measured the brightness of the rings from many angles as the spacecraft flew around the planet. They also measured changes in radio signals received as the craft passed behind the rings. If all the particles were compressed to form a single body, the body would be only about 100 km across. Data from the Voyager spacecraft confirmed that the particles consist of ice and ice-coated rocks.

Voyager also identified two tiny satellites, Prometheus and Pandora, each measuring about 50 km across, orbiting Saturn on either side of the F ring. They help to keep the icy particles into a well-defined, narrow band, about 100 km wide. The F ring also contains ringlets that are sometimes braided and sometimes separate. Dark radial 'spokes' that appear and disappear in the B ring are thought to be caused by Saturn's magnetic field.

The rings are thought to have formed from a cloud of particles that came from the breakup of a moon or from material that did not combine to form a moon. Moons that orbit within the rings act as **shepherd satellites** to create sharp edges and gaps between the rings.

DID YOU KNOW?

Instruments on board the Cassini space probe, which currently orbits Saturn, have detected an eruption of atomic oxygen in Saturn's E ring. The eruption may have been caused by objects colliding with particles in the ring or perhaps from a meteorite crashing into the ring. Such eruptions may indicate the ring is slowly eroding, and might even disappear within the next 100 million years. Scientists also think that mutual collisions between the particles in the rings will rob them of energy and they will over time spiral into Saturn.

Observations made by the Cassini probe in 2006 showed that the D ring isn't flat like the other rings. It appears to have corrugations like a tin roof. These corrugations are thought to have been caused by an impact as recently as 1984. In September 2006, Cassini also discovered two new diffuse rings made of tiny dust grains. One of them, R/2006S1, lies at the same distance as the co-orbital moons Janus and Epimetheus, 151 500 km from the planet's centre. The other, R/2006S2, overlies the orbit of the tiny moon Pallene at a radial distance of 212 000 km. Two more rings, R/2006S3 and R/2006S4, were found inside the dark, narrow Cassini division.

TEMPERATURE AND SEASONS

At the cloud tops and in the rings, the temperature is about −185°C. Frozen water is in no danger of melting or evaporating at these low temperatures. Temperature and pressure increase with depth below the cloud tops. In the outer core, the temperature reaches about 12 000°C and the pressure is about 12 million times the pressure on Earth's surface.

Because Saturn takes 29.46 years to orbit the Sun, any season on Saturn would last more than seven Earth years.

MAGNETIC FIELD

Saturn's magnetic field was first detected with the fly-by of NASA's Pioneer 11 spacecraft in 1979. Convection currents in the mantle of liquid metallic hydrogen generate the strong magnetic field. Saturn's field is about 36 times less powerful than the field of Jupiter but 540 times more powerful than Earth's field. Because the magnetic field is less powerful than Jupiter's, fewer charged particles are trapped in Saturn's magnetic field. The rings and moons also absorb some charged particles.

Saturn's magnetosphere is intermediate in size between Earth and Jupiter's, but it extends beyond the orbit of Saturn's moon Titan. Data from space probes show that Saturn's magnetosphere contains radiation belts similar to those of Earth. Variations in the magnetic field are thought to be responsible for the presence of dark spokes seen moving in Saturn's rings.

On Earth, the magnetic polar axis and the rotational axis vary by about 11°, but on Saturn the two axes are within 1° of each other.

In June 2005 the Cassini spacecraft detected auroral emissions around both poles of Saturn. The blue–ultraviolet emissions are thought to be caused by hydrogen gas being excited by electron bombardment. The

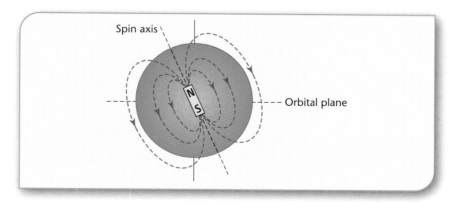

Figure 10.7   Saturn's magnetic field.

images showed that the auroral light responded rapidly to changes in the solar wind strength.

MOONS OF SATURN

Saturn has 62 moons, nearly as many as Jupiter. The moons fall into three groups: there are 21 moons between 133 000 km and 527 000 km from the planet, three between 1 221 000 km and 3 560 000 km, and the rest between 11 294 000 km and 24 500 000 km.

The moons range in diameter from about 3 km to 5150 km, but the smaller ones are more asteroid-like and may not be true moons. Some astronomers call them 'moonlets'. Most of these moons are icy worlds heavily covered with craters caused by impacts very long ago. Some astronomers believe that the moons may have condensed from a series of gas rings cast off from Saturn about 4.5 billion years ago.

There were 18 known moons orbiting Saturn when the Cassini spacecraft began its journey to the planet in 1997. During Cassini's seven-year journey to Saturn, Earth-based telescopes discovered 13 additional moons. When

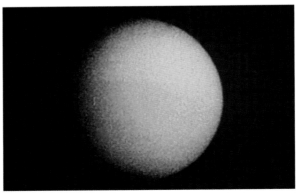

Figure 10.8   Saturn's largest moon is Titan. It is the only moon in the solar system with a substantial atmosphere. In 2004 the Huygens probe took this photograph of the surface of Titan while descending through Titan's atmosphere. (Photo: NASA)

Cassini reached Saturn in 2004, three more moons were discovered (Methone, Pallene and Polydeuces). On 1 May 2005, Cassini found another moon, hidden in a gap in Saturn's outer A ring.

In 2006, astronomers using the 8.2-metre Subaru Telescope in Hawaii detected nine more moons orbiting Saturn. These moons or satellites are about 6 to 8 kilometres in size and they travel on highly eccentric, retrograde orbits (opposite to the planet's rotation). These objects were probably captured by Saturn's gravity.

As of mid-2008, a total of 62 moons or satellites have been detected around Saturn.

Many of the moons have been officially named and the rest have been given temporary numbers until they are fully confirmed.

Table 10.3 Details of the moons of Saturn

Name	Distance from Saturn (km)	Period (days)	Diameter (km)	Discovered
Pan	134 000	0.58	20	1990
Daphnis	136 000	0.59	7	2005
Atlas	137 000	0.60	32	1980
Prometheus	139 000	0.62	100	1980
S/2004S4	140 100	0.62	4	2004
S/2004S6	140 130	0.61	4	2004
S/2004S3	140 300	0.62	4	2004
Pandora	141 700	0.63	84	1980
Epimetheus	151 400	0.69	116	1980
Janus	151 500	0.70	178	1966
Mimas	185 520	0.94	397	1789
Methone	194 000	1.01	3	2004
Pallene	212 000	1.15	4	2004
Enceladus	238 000	1.37	500	1789
Tethys	294 660	1.89	1060	1684
Telesto	294 700	1.89	24	1980
Calypso	294 700	1.89	19	1980
Dione	377 400	2.74	1120	1684
Helene	377 400	2.74	32	1980
Polydeuces	377 400	2.74	4	2004
Rhea	527 100	4.52	1530	1672
Titan	1 221 850	15.95	5150	1655

Name	Distance from Saturn (km)	Period (days)	Diameter (km)	Discovered
Hyperion	1 464 000	21.28	266	1848
Iapetus	3 560 800	79.33	1440	1671
Kiviuq	11 300 000	448	14	2000
Ijiraq	11 400 000	451	10	2000
Phoebe	12 900 000	545	220	1898
Paaliaq	15 103 000	692	20	2000
Skathi	15 672 000	732	8	2000
Albiorix	16 300 000	774	30	2000
Anthe	16 560 000	793	6	2007
Bebhionn	17 150 000	838	6	2004
Erriapo	17 236 000	844	10	2000
Skoll	17 473 800	862	6	2006
Siarnaq	17 777 000	884	40	2000
Tarqeq	17 910 600	895	7	2007
S/2004S13	18 056 300	906	6	2004
Greip	18 065 700	906	6	2006
Hyrokkin	18 168 300	914	8	2004
Jarnsaxa	18 557 000	944	6	2006
Tarvos	18 560 000	944	15	2000
Mundilfari	18 725 000	956	7	2000
S/2006S1	18 930 000	972	6	2006
S/2004S17	19 090 000	986	4	2004
Bergelmir	19 104 000	985	6	2004
Narvi	19 395 000	1008	7	2003
Suttungr	19 580 000	1022	7	2000
Hati	19 709 000	1033	6	2004
S/2004S12	19 905 000	1048	5	2004
Farbauti	19 984 000	1054	5	2004
Thrymr	20 278 000	1078	7	2000
Aegir	20 482 000	1094	6	2004
S/2007S3	20 518 000	1100	5	2007
Bestla	20 570 000	1100	7	2004
S/2004S7	20 576 000	1102	6	2004
S/2006S3	21 076 000	1142	6	2006
Fenrir	21 930 000	1212	4	2004
Surtur	22 288 000	1242	6	2006
Kari	22 321 000	1245	7	2006
Ymir	22 430 000	1254	16	2000
Loge	22 984 000	1300	6	2006
Fornjot	24 500 000	1300	6	2004

Did You Know?

Astronomers keep finding new moons around Saturn using both the Cassini probe and ground-based telescopes. The rings of Saturn make observation of moons difficult because the rings tend to hide some of the moons. Some astronomers believe the rings may have formed from dust material being pulled off the surface of the moons by Saturn's gravity. Other astronomers believe the opposite, that is, that the small moons may have formed or are still forming from condensed ring material. A number of the smaller moons (or moonlets) are shaped like a flying saucer. It is thought that these moons each collected material from the rings to grow around their middles. In 2008, Cassini scientists reported that material is continually on the move both into and out of the moonlets.

Only seven of the known moons of Saturn are massive enough to have collapsed into a spherical shape. The rest are irregular in shape, suggesting they are captured asteroids. Unlike Jupiter, Saturn has only one planet-sized moon; this is called **Titan**. Titan is second in size only to Ganymede among the moons in the solar system, and it is also larger than the planet Mercury. It can be seen fairly easily through telescopes from Earth.

Christiaan Huygens discovered Titan in 1655, the same year he discovered the rings. It has a diameter of 5150 kilometres and it orbits Saturn at a distance of about 1.2 million kilometres. Titan takes about 16 days to orbit Saturn and it also takes this time to rotate once on its axis. Thus it always has the same side facing Saturn.

Titan is thought to be made of half water ice and half rock or silicates. Its rocky core is surrounded by a mantle of ice and an icy crust that may contain some liquid water. Titan is the only moon in the solar system to have a dense atmosphere. Brown-orange clouds completely obscure its surface, and little sunlight reaches the surface. Voyager data showed most of the atmosphere is nitrogen gas (94%), with the rest mainly methane (5%). The nitrogen gas may have originally been in the form of ammonia, which broke up into hydrogen and nitrogen. Hydrogen, being light, may have escaped Titan's weak gravity. There are traces of hydrocarbons, such as methane, acetylene, ethane, ethylene and propane. All the oxygen is present as water ice. The atmosphere is four times as dense as Earth's, but because of the weak gravity atmospheric pressure is only 1.6 times greater than Earth's.

The temperature on Titan is about $-178°C$, which is near the freezing point of methane. Thus it is expected that the surface contains lakes of hydrocarbon liquids like ethane. Nitrogen reacts with these hydrocarbons to produce other compounds, some of which are the building blocks of organic molecules essential for life. In 1994 the Hubble Space Telescope observed bright and dark areas on Titan that may be impact craters, oceans or continents.

Initial pictures taken by the Cassini spacecraft in July 2004 showed a murky landscape with a variety of features, such as giant equatorial sand dunes, polar lakes, and methane-soaked mud flats. So far, only one mountain ridge has been detected. There is evidence of volcanoes, flows and calderas. The northern region contains well-defined lakes, channels and islands. The first infrared pictures revealed water ice as dark patches and masses of clouds in the southern hemisphere.

Cassini also mapped the interaction between the huge magnetosphere that surrounds the Saturn system and Titan's dynamic atmosphere. The 80 000 km-wide gas cloud that follows Titan as it orbits the planet is evidence that its atmosphere is breaking up.

On 24 December 2004, the Huygens probe was detached from Cassini and headed towards Titan. The probe is about the size of a small car and shaped like a flying saucer. On 14 January 2005, the Huygens probe entered Titan's atmosphere at about 6 km/s, and its heat shield reached $8000°C$. Three parachutes were used to slow the probe down and it landed on the surface with a 'splat' in what appeared to be mud. The first images of the surface showed a pale orange, eroded landscape of rocks and ice blocks, together with what looked like drainage channels. A thin crust gives the surface a squishy consistency.

OTHER MOONS OF SATURN

Excluding Titan, the six largest moons range in diameter from 397 km to 1530 km. In order of increasing distance from Saturn, these moons are **Mimas, Enceladus, Tethys, Dione, Rhea** and **Iapetus**. These moons have average densities around 1.2 g/cm^3, indicating they are made mostly of water ice with some rock. Having formed in a cold environment, these bodies retained water, methane, ammonia and nitrogen that condensed from the solar nebula. Some astronomers believe these mid-sized moons condensed from gas rings that surrounded Saturn about 4.5 billion years ago.

Mimas has a diameter of 397 km and is 185 520 km from Saturn. It takes only 23 hours to orbit Saturn. Mimas was discovered in 1789 by William Herschel. It is difficult to observe from Earth because it is so close to

Figure 10.9   Dione, one of Saturn's moons. (Photo: NASA)

Saturn. The Voyager probe showed that the surface of Mimas is dominated by a large impact crater, called Herschel, which is 130 km across, one-third the diameter of Mimas. The walls of this crater are about 5 km above the surface and parts of its floor are 10 km deep. It has a central peak that rises 6 km above the crater floor. Fractures can be seen on the opposite side of Mimas that may be due to the large impact. Other smaller impact craters also exist on Mimas.

Enceladus has a diameter of 500 km and is 238 000 km from Saturn. Herschel discovered it in 1789. The surface of this moon is covered with a smooth layer of water ice that makes it the most reflective of any known planetary body. There are many craters in one hemisphere and very few in the other hemisphere. The young surface of this moon was seen by Voyager 2 to contain a number of different types of formations, including ice flows, faults and striations. The crust is probably thin and lying on top of a molten interior. Ice volcanoes on Enceladus are a likely source for the particles that create Saturn's outermost E ring.

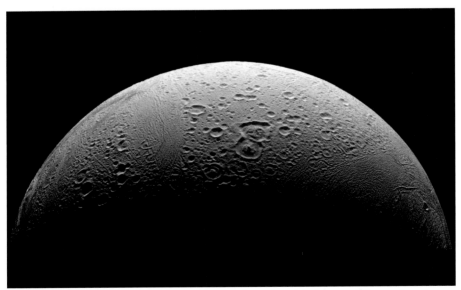

Figure 10.10   Surface of the moon Enceladus taken by Cassini. The surface is mostly cratered and crossed by numerous fault lines. (Photo: NASA)

DID YOU KNOW?

During 2006, it was reported that the Cassini spacecraft had found evidence of water spewing from geysers on the Saturnian moon called Enceladus. As a result some scientists have now placed this moon on the short-list of places most likely to have extraterrestrial life. High-resolution images snapped by the orbiting Cassini probe confirmed the eruption of icy jets and giant water plumes from geysers at Enceladus' south pole. If any life exists on the moon, it probably would be in the form of microbes or other primitive organisms.

Tethys has a diameter of 1060 km and is 294 660 km from Saturn. Giovanni Cassini discovered it in 1684. Like Mimas, the surface of this moon is heavily cratered in places, but there are also some smooth areas. The largest impact crater is Odysseus, with a diameter of 400 km (making it larger than the moon Mimas). This crater is relatively shallow, unlike the crater on Mimas. A valley called Ithaca Chasma stretches three-quarters the way around the moon and in places is up to 100 km wide. The Chasma may have been formed by the impact or by tectonic activity.

Dione is 377 400 km from Saturn and has a diameter of 1120 km (Figure 10.9). It, too, is heavily cratered, with a number of large impact craters scattered over its surface. There are also some smooth areas that may be coverings of water ice. Recent images show the surface contains features such as faults, valleys and depressions, caused by tectonic movement. The largest craters are about 100 km across, and bright streaks that are seen radiating from some craters are the result of material being ejected after impact.

Rhea is Saturn's second largest moon, being 1530 km in diameter. This moon is 527 100 km from Saturn. The surface is very old and shows little change by geological activity. Like Dione, parts of its surface are heavily cratered and parts are smooth. There are also some white streaks across the surface that may be ice-filled cracks. Rhea is called the 'ammonia moon' by some astronomers because it consists of about 25 per cent frozen ammonia in addition to rock and water ice.

Iapetus is the third largest moon orbiting Saturn. It has a diameter of 1440 km and orbits Saturn at a distance of 3 560 800 km from Saturn. This moon was discovered by Giovanni Cassini in 1671. One side of this moon is cratered and bright (probably water ice), while the other side is coated with a dark (probably carbon-based) material of unknown origin. The dividing line between the two regions is relatively sharp. The Cassini probe revealed Iapetus to be heavily cratered with a bizarre patchwork of pitch-dark and snowy-white regions. High-resolution images show a 20-kilometre-high ridge running along part of its equator.

Figure 10.11   Close up view of the cratered surface of the moon Iapetus.
(Photo: NASA)

Most of the remaining moons of Saturn range in diameter from 3 km to 266 km. Many of these moons are irregular in shape and have unusual orbits, suggesting they are fragments of larger bodies or captured asteroids.

Hyperion is the largest of Saturn's minor moons, while **Phoebe** orbits in a direction opposite to the orbits of the other moons and opposite to the direction of Saturn's rotation. Phoebe also has a highly inclined orbit and its surface is covered with a dark material (Figure 10.13).

Pan is the innermost of Saturn's known moons, orbiting at a distance of 134 000 km. With a diameter of only 20 km, it was discovered from Voyager photographs in 1990. Pan orbits within the Encke division in Saturn's A ring. Small moons near the rings produce wave patterns in the rings.

Janus and **Epimetheus** are two irregularly shaped moons that are co-orbital (orbit together). These two bodies are separated by less than 100 km and their velocities are nearly equal. Their gravitational interaction causes them to exchange orbits every four years. Astronomers believe they are probably fragments of a single body, now destroyed.

Figure 10.12   The Cassini space probe took this photograph of Epimetheus in December 2007. The moon is 116 km in diameter and irregular in shape. Heavy cratering on its surface indicates it may be several billion years old. (Photo: NASA)

Figure 10.13 The moon Phoebe has a diameter of 220 km and orbits in a direction opposite to that of the other moons. (Photo: NASA)

WEB NOTES

For fact sheets on any of the planets, including Saturn, check out:
    <http://nssdc.gsfc.nasa.gov/planetary/planetfact.html>
    <http://www.space.com/saturn/>
    <http://www.nasm.si.edu/etp/>

Chapter 11

Uranus –
The Blue, Icy Planet

Uranus is the third of the gas giants and the seventh planet from the Sun. The planet is the third largest in the solar system, with a diameter of 51 118 km. Uranus is one-third the diameter of Jupiter but four times the diameter of Earth. It is large enough to hold about 64 Earths. Its distance

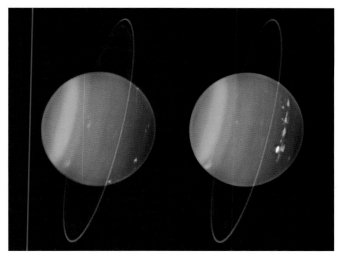

Figure 11.1 The gas giant Uranus has a faint ring system around it. These two IR images were taken half a Uranian day apart by the Keck Telescope in Hawaii. The white patches are clouds. (Photo: W.M. Keck Observatory)

from the Sun varies from about 3010 million kilometres at its furthest point to about 2739 million kilometres at its closest point. Because of this large distance, it takes Uranus 84 years to orbit the Sun once. Since its discovery in March 1781 it has gone around the Sun only just over two and half times. Light from the Sun takes just over 8 minutes to reach Earth, but it takes about 2 hours 40 minutes to reach Uranus.

Uranus travels around the Sun in a slightly elliptical orbit and it spins on its axis once every 17 hours 14 minutes. It is the furthest planet that can be seen from Earth with the unaided eye; however, it is faint and difficult to detect. Uranus was the first planet to be discovered by telescope.

Uranus is pale blue in colour with few surface features. It has a faint ring system and an extensive family of moons, but little was known about the planet until the Voyager 2 probe flew by it in January 1986.

EARLY VIEWS ABOUT URANUS

Uranus is the ancient Greek deity of the heavens, the earliest supreme god. Uranus was the son and mate of Gaia and the father of Cronos (Saturn) and of the Cyclopes and Titans (predecessors of the Olympian gods). Uranus is diffcult to see in the night sky because it is so far away from the Sun and therefore faint. The first recorded sighting of this object was made in 1690 by the English astronomer John Flamsteed, who incorrectly catalogued it as 34 Tauri. Another observer, Pierre Lemonnier, recorded Uranus as a star a total of 12 times during 1769.

William Herschel officially discovered Uranus in 1781 while observing the night sky through one of his telescopes. Herschel noted that the object moved against the background of stars over several nights and had a bluish-green disc, unlike the stars, which were point sources of light. Herschel concluded that the faint object was a planet and he called it 'the Georgium Sidus' (the Georgian planet) in honour of his patron, King George III of England; other people called the planet 'Herschel'. At that time Saturn was the furthest known planet and its discovery effectively doubled the size

Table 11.1 Details of Uranus

Distance from Sun	2 870 990 000 km (19.2 AU)
Diameter	51 120 km
Mass	8.68×10^{25} kg (14.53 times Earth's mass)
Density	1.29 g/cm^3 or 1290 kg/m^3
Orbital eccentricity	0.046
Period of revolution (length of year)	30 685 Earth days or 84.01 Earth years
Rotation period	17 hours 14 minutes
Orbital velocity	24 516 km/h
Tilt of axis	97.86°
Average temperature	–200°C
Number of moons	at least 27
Atmosphere	hydrogen, helium, methane
Strength of gravity	8.2 N/kg at surface

of the known solar system. Herschel also discovered the two largest moons of Uranus, Oberon and Titania.

Johann Bode, a German astronomer, named the planet Uranus after the Graeco-Roman god who personified the universe and was the father of Saturn. The planet was officially named Uranus in 1850.

Probing Uranus

We know little about Uranus because it is so far from Earth. Most of what we do know came from the Voyager 2 probe, which flew by Uranus in early 1986. The probe passed within 82 000 km of the planet's cloud tops. It took Voyager 2 five years to travel from Saturn to Uranus.

Photographs taken by Voyager 2 revealed that Uranus was a blue planet with a few faint bands of clouds moving parallel to its equator. There were no signs of belts or storm spots. Ten additional moons were discovered around the planet, and most have at least one shattering impact crater. Voyager 2 also found that the planet's magnetic field was 50 times stronger than Earth's field. The magnetic field is tilted at 59° to its axis of rotation, and does not pass through the centre of the planet. Because of this strange arrangement, the magnetic field wobbles considerably as the planet rotates.

Table 11.2 Significant space probes to Uranus

Probe	Country of origin	Launched	Comments
Voyager 2	USA	1977	Passed by Uranus in January 1986

DID YOU KNOW?

The 20th anniversary of the closest approach of Voyager 2 to the gas giant Uranus was on 24 January 2006. William Herschel discovered the planet in 1781, but it remained a mysterious planet for 205 years with little known about this distant world.

Before the January 1986 fly-by of Voyager 2, planetary scientists relied on ground-based observational data and inference to determine the nature of the Uranian system. When the programmed fly-by of Uranus approached, scientists and ground-based controllers scrambled to prepare Voyager 2's onboard systems to capture images and analyse the planet.

POSITION AND ORBIT

Uranus is the third largest planetary member of the solar system and is the seventh planet from the Sun. Its orbit is slightly elliptical and lies between those of Saturn and Neptune. Uranus has a mean distance from the Sun of about 2871 million kilometres, placing it about 19.2 times further from the Sun than is Earth. It travels around the Sun once every 84.01 years, and rotates on its axis with a period of 17 hours 14 minutes. The planet's atmosphere rotates faster than its interior. The fastest winds on Uranus, measured about two-thirds of the way from the equator to the south pole, blow at about 720 km/h.

Most planets spin on an axis nearly perpendicular to their orbital plane, but Uranus' axis is almost parallel to this plane (nearly 98° to the vertical). This means the rotational axis is almost 8° below the orbital plane, so the

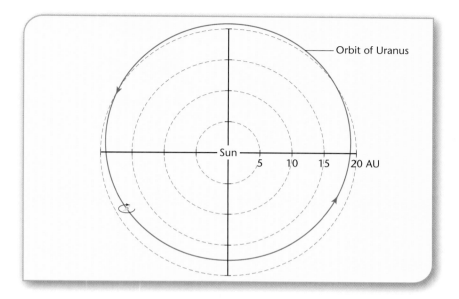

Figure 11.2 Orbital path of Uranus (distance circles are in astronomical units, AU).

planet appears to be tipped on its side. It is not known why Uranus has such a high axial tilt, but it may have been hit by another large body sometime in its past. At the time of Voyager 2's fly-by, the planet's south pole was pointed almost directly at the Sun. As a result of this, Uranus's polar regions receive more energy from the Sun than its equatorial regions. However, the temperature is still higher at the equator than at its poles, for unknown reasons.

Uranus's spin is also retrograde, meaning it is spinning in the opposite direction to most other planets.

DENSITY AND COMPOSITION

Uranus is a gaseous, icy planet with a mass about 14 times that of Earth, but only about one-twentieth that of the largest planet, Jupiter. The average

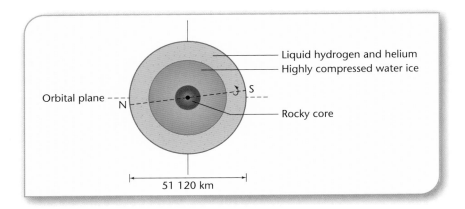

Figure 11.3   Interior structure of Uranus.

density of Uranus is about 1.3 g/cm³, about one-quarter that of Earth. Thus the material Uranus is made out of must be light and icy.

In contrast to the other gas planets (Jupiter and Saturn), the composition of Uranus is not dominated by hydrogen and helium. Hydrogen accounts for only 15 per cent of the planet's mass. Most of the planet is made up of methane, ammonia and water. There are three layers or regions inside the planet. The dense core (30%) contains rock and various ices, but no liquid metallic hydrogen. The mantle (40%) is probably highly compressed water ice with some methane and ammonia. The outer layer (30%) lies at the base of the atmosphere and is considered to be composed of mostly icy molecules of water, methane and ammonia.

The strength of gravity on Uranus is less than on Earth (8.2 N/kg compared to Earth's 9.8 N/kg). This means that a 75 kg person who weighs 735 N on Earth would weigh only 615 N on Uranus.

THE SURFACE

Being a gaseous planet, there is no solid surface layer on Uranus. The outer layer of the planet is probably made up of icy molecules of of water, methane and ammonia. The surface we see from Earth is Uranus's atmosphere. Clouds are visible in the atmosphere. The planet radiates about the same amount of energy as it receives from the Sun and has little internal heat.

THE ATMOSPHERE

The atmosphere of Uranus is composed of 83 per cent hydrogen, 15 per cent helium, 2 per cent methane, and tiny amounts of ethane and other hydrocarbon gases. The methane that is trapped high in the atmosphere absorbs red light from the visible spectrum, and this makes the planet appear blue-green in colour.

Voyager 2 data showed the atmosphere contains three distinct cloud layers. The top layer contains ammonia, the next layer ammonium hydrosulfide, and the third or lower layer contains water ice. These layers are found deep in Uranus's atmosphere, where temperatures and pressures are higher. The atmospheric pressure beneath the cloud layer is about 1.3 times that at the Earth's surface.

Like the other gas planets, Uranus has bands of clouds that blow around rapidly parallel to the equator. These bands are very faint and can only be

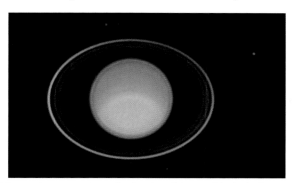

Figure 11.4   Uranus's atmosphere lacks detail in visible light but colour enhancement shows a dark polar region surrounded by a series of lighter concentric bands that may be smog or haze (Voyager 2). (Photo: NASA)

seen with image enhancement of the Voyager 2 photographs. Winds at mid-latitudes are propelled in the rotational direction of the planet. Winds at equatorial latitudes blow in the opposite direction.

Recent observations with the Hubble Space Telescope show larger and more pronounced streaks and some spots. The spots are probably violent swirling storms like a hurricane.

THE RINGS

Uranus has a number of rings around it. These were discovered by chance in 1977 when Uranus appeared to pass in front of the faint star SAO158687 in Libra, as seen from Earth. Such an event is called an **occultation**. The rings temporarily interrupted light from the star and pulses of starlight were seen each side of the planet, suggesting there was something around the planet. In 1986, Voyager 2 confirmed the existence of a ring system containing nine main rings with two smaller ringlets. The ring system lies in the Uranian equatorial plane, circling Uranus between 38 000 km and 51 140 km from its centre. Their overall diameter is over 100 000 km.

The rings are faint and composed of particles ranging in size from fine dust to several metres in diameter. Voyager 2 found that the gaps between

Table 11.3 Uranus' rings

Ring	Distance (km)	Width (km)
1986 U2R	38 000	2500
6	41 840	1–3
5	42 230	2–3
4	42 580	2–3
Alpha	44 720	7–12
Beta	45 670	7–12
Eta	47 190	0–2
Gamma	47 630	1–4
Delta	48 290	3–9
1986 U1R	50 020	1–2
Epsilon	51 140	20–100

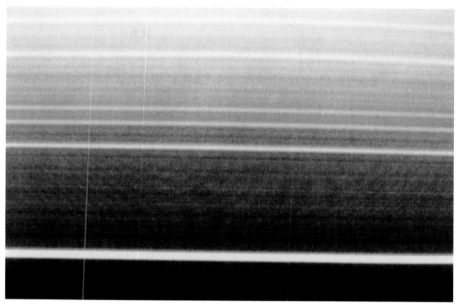

Figure 11.5 Voyager 2 image of the rings of Uranus showing slight colour differences. The bright white ring at the bottom is Epsilon. (Photo: NASA)

the rings were not empty but contained fine dust. This may have originated from collisions between the larger particles that form the main rings, or from the surrounding moons.

The outer ring is the most massive, and its particles are kept in orbit by the gravitational influence of two moons, Cordelia and Ophelia. The outermost ring is about 100 km wide but only 10 m to 100 m thick. The ring material is probably composed of chunks of ice, covered by a layer of carbon.

The Uranian rings were the first rings to be discovered after Saturn's. This was an important finding, since we now know that rings are a common feature of planets, not a peculiarity of Saturn alone. Uranus's rings are much darker than those of Saturn and are much harder to see from Earth.

DID YOU KNOW?

The outermost ring of Uranus, discovered in 2005, has been found to be bright blue. This ring is only the second known blue ring in the solar system – the outer ring of Saturn is also blue. Astronomers suspect both rings owe their colour to forces acting on dust in the rings that allow smaller particles to survive while larger ones are recaptured by nearby moons. Hubble discovered a new moon, now called Mab, orbiting Uranus in the same region as its blue ring. Mab is a dead rocky world about 24 km in diameter. Meteoroid impacts continually blast dust off the surface of Mab and this may contribute to the material in the blue ring of Uranus.

TEMPERATURE AND SEASONS

The temperature in the upper atmosphere of Uranus is about −200°C. At this low temperature, methane and water condense to form clouds of ice crystals. Because methane freezes at a lower temperature than water, it

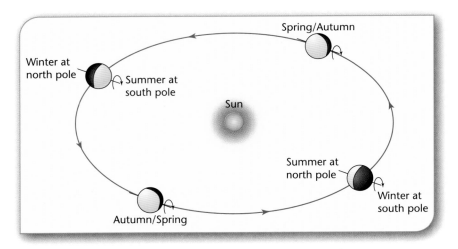

Figure 11.6   Exaggerated seasons on Uranus.

forms higher clouds over Uranus. Methane absorbs red light, giving Uranus its blue-green colour.

In the interior, the temperature rises rapidly to about 2300°C in the mantle and about 7000°C in the rocky core. The pressure in the core is about 20 million times that of the atmosphere at Earth's surface.

The planet radiates back into space as much heat as it receives from the Sun. Because its axis is tilted at 98°, its poles receive more sunlight during a Uranian year than does its equator. However, the weather system seems to distribute heat fairly evenly over the planet.

As the planet orbits the Sun, its north and south poles alternately point directly towards or directly away from the Sun, resulting in exaggerated seasons. During summer near the north pole, the Sun is almost directly overhead for many Earth years. At the same time southern latitudes are subjected to a continuous frigid winter night. Forty-two years later, the situation is reversed.

In August 2006, the Hubble Space Telescope captured images of a huge dark cloud on Uranus. The cloud measured about 1700 km by 3000 km. Scientists are not certain about the origin of the cloud; nor do they now how long it will last.

MAGNETIC FIELD

Uranus has a magnetic field about 50 times stronger than Earth's. The axis of the field (an imaginary line joining its north and south poles) is tilted 59° from the planet's axis of rotation. The centre of the magnetic field does not coincide with the centre of the planet – it is offset by almost one-third of Uranus's radius, or nearly 7700 km. Because of the large angle between the magnetic field and its rotation axis, the magnetosphere of Uranus wobbles considerably as the planet rotates.

Voyager 2 passed through Uranus's magnetosphere as it flew by Uranus. The magnetic field traps high-energy, electrically charged particles, mostly

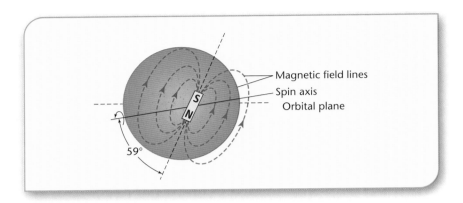

Figure 11.7   Uranus's strange magnetic field.

electrons and protons, in radiation belts that circle the planet. As these particles travel back and forth between the magnetic poles, they emit radio waves. Voyager 2 detected these waves, but they are weak and cannot be detected from Earth.

Uranus's magneto tail as measured by Voyager 2 was found to extend 10 million kilometres out into space. Unlike the magneto tails of other planets, the magnetic field lines in Uranus's tail are cylindrical and appear wound around each other like a corkscrew. This is probably due to the strange axial tilt of the planet.

MOONS

Before the visit of Voyager 2 to Uranus in January 1986, only five moons were known to orbit Uranus. These moons were discovered between 1787 and 1948 and they range in size from 471 km to 1578 km in diameter. Voyager discovered 10 more moons, all less than 50 km in diameter. Several of these tiny, irregularly shaped moons are shepherd satellites whose gravitational pull confines the rings of Uranus. Astronomers used Earth-based telescopes to find two more moons in 1997, and three more in 1999.

Table 11.4 Moons of Uranus

Name of moon	Distance (km)	Period (days)	Diameter (km)	Discovered
Cordelia	49 800	0.34	40	1986
Ophelia	54 000	0.38	42	1986
Bianca	59 200	0.44	51	1986
Cressida	62 000	0.46	80	1986
Desdemona	63 000	0.47	64	1986
Juliet	64 400	0.49	93	1986
Portia	66 000	0.51	135	1986
Rosalind	70 000	0.56	72	1986
Cupid	74 800	0.60	10	2003
Belinda	75 000	0.62	80	1986
Perdita	76 400	0.64	20	1986
Puck	86 000	0.76	162	1986
Mab	97 700	0.90	10	2003
Miranda	130 000	1.41	471	1948
Ariel	191 000	2.52	1157	1851
Umbriel	266 000	4.14	1170	1851
Titania	436 000	8.71	1578	1787
Oberon	583 000	13.46	1522	1787
Francisco	4 276 000	266	12	2001
Caliban	7 230 000	579	98	1997
Stephano	8 004 000	677	20	1999
Trinculo	8 504 000	749	10	2001
Sycorax	12 180 000	1288	190	1997
Margaret	14 345 000	1687	11	2003
Prospero	16 256 000	1978	30	1999
Setebos	17 418 000	2225	30	1999
Ferdinand	20 901 000	2887	12	2001

Unlike the other bodies in solar system that have names from classical mythology, the names of Uranus's moons are derived from the writings of Shakespeare and the English writer Alexander Pope.

Uranus has at least 27 moons, arranged in three groups: 13 small dark inner ones, five large moons, and nine more distant ones recently discovered by telescope. Most of these moons have nearly circular orbits in the plane of Uranus's equator. The outer moons have elliptical orbits and many have retrograde motion. It may be that some of the smaller moons, especially the outer ones, are captured asteroid-like bodies.

Figure 11.8   Voyager 2 photograph of the surface of Titania, the largest moon of Uranus. (Photo: NASA)

Most of the moons of Uranus are quite dark. This may be due to radiation darkening methane on their surfaces.

The five largest moons of Uranus (Miranda, Ariel, Umbriel, Titania and Oberon) have higher densities than expected (1.4 to 1.7 g/cm^3). This suggests that they may contain more than 50 per cent rock or silicates, with smaller amounts of water ice than Saturn's similar-sized moons.

Miranda has a surface like no other in the solar system. The older areas are relatively smooth, cratered plains, but other areas look as if they have been clawed and gouged by something. Astronomers think Miranda's core

Figure 11.9   A Voyager 2 photograph of the surface of Miranda. (Photo: NASA)

originally consisted of dense rock while its outer layers were mostly ice; however, this structure has at some time in the past been dramatically changed. Surface variations suggest an asteroid or tectonic activity may have shattered the moon, breaking its surface into several pieces – the pieces have since been reassembled by gravity, leaving behind great scars.

Ariel is a moon with few craters, most being less than 50 km in diameter. The surface suggests there has been a lot of geological activity, with many faults, fractures and valleys visible. Photographs taken by Voyager 2 suggest that many of the features are volcanic in origin and that some form of viscous fluid once flowed across the surface.

Umbriel is the darkest of the five larger moons orbiting Uranus. Its surface is almost uniformly covered with craters. Many large craters suggest the surface is fairly old, although the covering of dark material appears to hide many features.

Titania is the largest of the Uranian moons, with a diameter of 1578 km. The surface of this moon has numerous impact craters, with some young

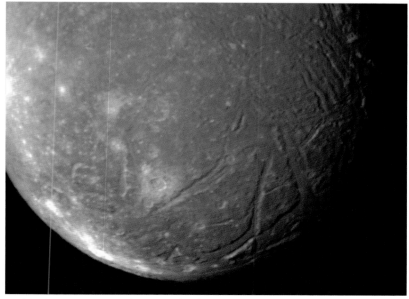

Figure 11.10   Voyager 2 photograph of the surface of Ariel. (Photo: NASA)

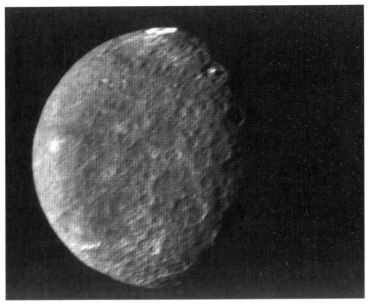

Figure 11.11   Oberon is Uranus's second largest moon. (Photo: NASA)

valleys, faults and fractures. One heavily fractured region is thought to have been caused by the crust fracturing as water froze below the surface.

Oberon has a diameter of 1522 km, slightly smaller than Titania. Oberon's surface has many impact craters, some of which are surrounded by bright rays. One mountain towers above the surrounding terrain to a height of 20 km.

WEB NOTES

For fact sheets on any of the planets, including Uranus, check out:
<http://nssdc.gsfc.nasa.gov/planetary/planetfact.html>
<http://www.space.com/uranus/>
<http://www.nasm.si.edu/etp/>

CHAPTER 12

NEPTUNE –
AN ICY, COLD WORLD

NEPTUNE IS THE smallest of the gas giants and the eighth planet from the Sun. It is the fourth largest planet in the solar system, with a diameter of 49 532 km. Neptune is smaller in diameter than Uranus but it has more mass than Uranus. Although four times the diameter of Earth, Neptune has only 17 times the mass of Earth because it is less dense than Earth.

Figure 12.1 Neptune as seen by Voyager 2 from a distance of 15 million kilometres. (Photo: NASA)

Neptune is about 4.5 billion kilometres from the Sun, making it about 1.5 times more distant than Uranus. It takes over 165 Earth years to orbit the Sun once, and it rotates on its axis once every 16.1 hours. Because of its great distance from the Sun, Neptune receives only 0.1 per cent of the sunlight that Earth receives, and the planet's surface is very cold.

The planet cannot be seen with the unaided eye from Earth. Neptune can be seen with good binoculars from Earth, but a large telescope is needed to see the tiny disc. It then appears as a tiny featureless disc that is barely distinguishable from a star. Most of our information about Neptune came from the Voyager 2 space probe that passed by Neptune in 1989. Voyager found Neptune to have a deep blue colour with an outer layer covered with whitish clouds.

EARLY VIEWS ABOUT NEPTUNE

In Roman mythology Neptune was the god of the sea.

Early astronomers did not know about Neptune as a planet because it could not be seen from Earth by the unaided eye. The discovery of Neptune was a great triumph of mathematics. After the discovery of Uranus in 1781, astronomers noticed that its orbit was not following the path predicted by Newton's and Kepler's laws of motion. Either these laws were wrong or there had to be another undiscovered planetary body in the solar system influencing Uranus's orbit. In 1845 and 1846, John Couch Adams in England and Urbain Le Verrier in France independently calculated the mass and location of such a body. The body, now called Neptune, was discovered in 1846 by German astronomer Johann Gottfried Galle, close to the predicted position. Within a few weeks the British astronomer William Lassell discovered a moon around Neptune, now named Triton after the half-man, half-fish son of the sea god.

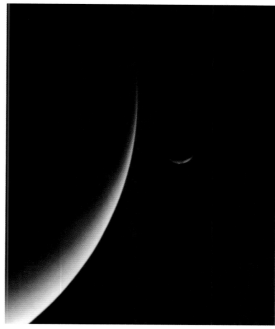

Figure 12.2   This dramatic view of the crescent of Neptune and its largest moon Triton was acquired by Voyager 2 when it was about 5 million kilometres from Neptune. (Photo: NASA)

Table 12.1 Details of Neptune

Distance from Sun	4 504 000 000 km (30.1 AU)
Diameter	49 532 km
Mass	1.02×10^{26} kg (17.1 times Earth's mass)
Density	1.64 g/cm³ or 1640 kg/m³
Orbital eccentricity	0.010
Period of revolution (length of year)	60 225 Earth days or 165 Earth years
Rotation period	16 hours 7 minutes
Orbital velocity	19 548 km/h
Tilt of axis	29°
Average temperature	–225°C
Number of moons	13
Atmosphere	hydrogen, helium, some methane
Strength of gravity	11.2 N/kg at surface

PROBING NEPTUNE

Astronomers know little about Neptune because it is so far from Earth. Neptune has been visited by only one spacecraft, **Voyager 2**, on 25 August 1989. Almost everything we know about Neptune has come from this encounter. Voyager had been travelling for about 12 years and had covered nearly 5 billion kilometres to reach Neptune. The space probe came to within 5000 km of the planet and it collected a wealth of information about this most distant gas giant and its moons.

Voyager found Neptune to be a large blue planet, with many markings and cloud bands. Five thin rings were also found around the planet, and six new moons were discovered to add to the two already known. The rings were found to be complete rings with bright clumps in them. One of the rings appeared to have a curious twisted structure. The rings are very dark and their composition is unknown.

Figure 12.3   Fluffy white clouds floating high in Neptune's atmosphere (Voyager 2). (Photo: NASA)

Table 12.2 Significant space probes to Neptune

Probe	Country of origin	Launched	Comments
Voyager 2	USA	1977	Passed by Neptune in August 1989

POSITION AND ORBIT

Neptune is the eighth planet from the Sun and the fourth largest planetary member of the solar system. Its orbit is slightly elliptical and lies beyond that of Uranus. Neptune has a mean distance from the Sun of about 4500 million kilometres, placing it about 30.1 times further from the Sun than is Earth. Neptune is so far away that when Voyager 2 was passing the planet the radio signals from the probe took over four hours to reach Earth.

Neptune travels around the Sun once every 165 years and rotates with a period of 16 hours 7 minutes. In its equatorial zone, winds blow westward at close to 1500 km/h, creating huge storms.

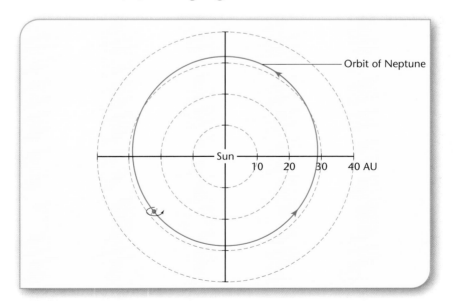

Figure 12.4 Orbital path of Neptune (distance circles are in astronomical units, AU).

Neptune spins on an axis at an angle of 29° from the vertical. This amount of axial tilt is similar to that of Earth. Because of this angle, it was expected that Neptune's poles would be colder than its equator. However, astronomers using Europe's Very Large Telescope in Chile found that the planet's south pole is about 10°C warmer than elsewhere. This imbalance in temperature probably explains why Neptune has such strong winds.

DENSITY AND COMPOSITION

Neptune is a gaseous planet with a mass about 17 times that of Earth, but it is not as dense as Earth. The average density of Neptune is about 1.64 g/cm³, compared to Earth's density of 5.52 g/cm³. This is mainly due to the different compositions of the two planets.

Neptune's composition and interior are similar to those of Uranus. Both planets have a rocky core surrounded by frozen ammonia, methane and water. Hydrogen contributes only about 15 per cent of the planet's total mass. Compared to Jupiter and Saturn, Neptune has more ammonia, methane and water.

Neptune's interior consists of a small but dense core of melted rock and ice. Because of Neptune's greater density, its core is probably slightly larger

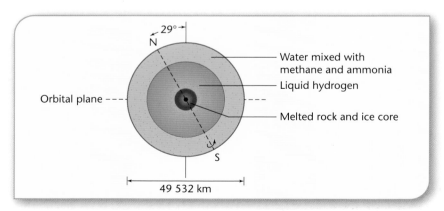

Figure 12.5   Interior structure of Neptune.

than the core of Uranus. Surrounding the core is a mantle of liquid hydrogen and an outer layer or ocean of water mixed with methane and ammonia. However, Neptune's interior layers may not be distinct and uniform. Like Jupiter and Saturn, Neptune produces its own heat – it radiates more than twice as much energy as it receives from the Sun.

The strength of gravity on Neptune is greater than Earth's gravity (11.2 N/kg compared to Earth's 9.8 N/kg). This means that a 75 kg person weighing 735 N on Earth would weigh 840 N on Neptune.

THE SURFACE

Being a gaseous planet, there is no solid surface layer on Neptune. Most of the planet consists of compressed, frozen gases. The outer layer of the planet is best described as an ocean containing water mixed with methane and ammonia. Thick clouds cover the ocean.

THE ATMOSPHERE

The atmosphere of Neptune contains a mixture of hydrogen, helium and some methane. Like the other gas giants, layers of ammonia, ammonium hydrosulfide and water ice are also thought to exist.

Neptune's atmosphere is much more active than that of Uranus. Rapid changes in weather occur regularly, and westward moving winds reach speeds up to 2000 km/h (the fastest winds of all the planets). The winds are driven by the heat energy radiated out from Neptune's interior.

As Voyager 2 passed within 5000 km of the Neptunian cloud tops, its cameras revealed a wide variety of features. Bright polar collars and broad bands in different shades of blue were prominent in the southern hemisphere. Also visible were bright streaks of cirrus cloud stretched out parallel to the equator. The Voyager pictures also detected shadows of these clouds thrown onto the main cloud layer some 50 km below.

Voyager 2 photographed large storms in the atmosphere. One in particular was about half the size of Jupiter's Great Red Spot, and oval in shape. Named the Great Dark Spot, it was observed to rotate anti-clockwise over a period of about 10 days. Observation revealed this feature was a hole in the Neptunian clouds through which the lower atmosphere could be seen. Winds blew the Spot westward at about 300 m/s. Cirrus-type clouds of frozen methane were seen forming and changing shape above and around the Great Dark Spot (Figure 12.6).

Voyager 2 also identified a smaller dark spot in the southern hemisphere and a small irregular white cloud that moves around Neptune every 16 hours or so, known as the 'Scooter'. This latter feature may be a gas plume rising from lower in the atmosphere.

Various white streaks and spots have been detected on Neptune from time to time, but most disappear or change greatly while other features

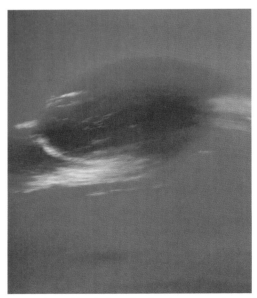

Figure 12.6 Neptune's Great Dark Spot had a diameter about the size of Earth. (Photo: NASA)

emerge. In 1994, images from the Hubble Space Telescope showed the Great Dark Spot had disappeared. (Compare Jupiter's Great Red Spot, which has lasted hundreds of years.) This indicates that Neptune's atmosphere changes rapidly.

Neptune is also covered with a number of belts and zones. These are fainter than those on Jupiter. A broad, darkish band is prominent at high southern latitudes. Embedded in this band is a smaller dark spot, about the size of Earth. White and wispy clouds seem to hover over the smaller dark spot. Scientists believe the darker clouds on Neptune contain hydrogen sulfide.

THE RINGS

Astronomers on Earth thought Neptune might have some incomplete rings or arcs when they observed an occultation of a star by Neptune in

Figure 12.7 Neptune's two main rings as seen by Voyager 2. The image of the planet is over-exposed to capture detail in the rings. (Photo: NASA)

1984. However, pictures taken by Voyager 2 revealed that these arcs were part of a narrow ring that contains three particularly bright areas. A second narrow ring was identified, as well as two other rings that were more spread out. The ring closest to Neptune is about 42 000 km above the cloud tops. The outer ring, which contains the arcs, is about 63 000 km from Neptune.

Neptune's rings are hard to see because they consist of small particles that reflect little light. Their exact composition is unknown but, because the temperature is so low, they probably contain frozen methane. Some of this methane has been changed by radiation into other carbon compounds, thus making the rings appear dark.

TEMPERATURE AND SEASONS

The large distance between Neptune and the Sun means that the average temperature on Neptune is a very cold –225°C. This temperature is low enough to freeze methane. Bright clouds of methane ice form in the upper atmosphere, and cast shadows on the lower cloud layers.

The planet's axis is tilted by nearly 29° from the vertical, but since Neptune takes 165 years to orbit the Sun the time between any seasons is very long and temperatures do not vary much season to season.

Neptune's interior is believed to be very similar in composition and structure to that of Uranus. Core temperature is therefore expected to be around 7000°C and core pressure about 20 000 atmospheres.

MAGNETIC FIELD

Neptune has a strong magnetic field, about 25 times greater than Earth's. However, the magnetic axis is tilted 47° from Neptune's rotational axis, and it is off-centre by more than half the radius of the planet. The magnetic

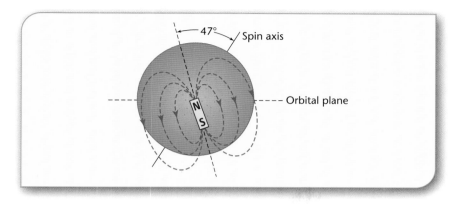

Figure 12.8 Neptune's magnetic field.

field is probably generated in middle regions where the pressure is high enough for water to conduct electrical currents.

MOONS OF NEPTUNE

Prior to the Voyager 2 encounter, only two moons were known, Triton and Nereid. Today we know Neptune has thirteen moons, following the discovery of six new moons by Voyager 2 and later observations. William Lassell discovered the largest moon, Triton, in 1846. Triton is spherical in shape and has a nearly circular orbit. The other moons are small, irregularly shaped objects with highly elliptical orbits, suggesting Neptune has captured them. One of the interesting things about Triton is that it has a retrograde orbit; that is, it orbits in the opposite direction to Neptune's rotation. Some astronomers believe Neptune may also have captured Triton some three to four billion years ago.

Triton has a diameter of 2700 km and orbits Neptune every 5.88 days at a distance of 354 800 km. Its surface contains many interesting features, including fault lines, cracks and ice (water, methane and ammonia) flows. There are not many impact craters, an indication that ice flows from the interior may have caused extensive resurfacing. The equatorial region has a wrinkled terrain that resembles the skin of a cantaloupe or rock melon. Long narrow valleys rimmed by ridges cross the area. Such a region may have formed from repeated episodes of melting and cooling of the icy crust. Triton also has a few frozen lakes that may be the calderas of extinct ice volcanoes. In other areas, dark features surrounded by bright aureoles or rings are visible – these may be some of the geyser-like plumes detected by Voyager 2. The pinkish South Polar Region is covered by a cap of frozen methane and nitrogen, and temperatures there are around –245°C. The pinkish colour of the polar cap is probably due to frozen nitrogen.

The very thin atmosphere of Triton contains nitrogen and methane.

Triton's density was found to be a little over 2 g/cm³, roughly twice the density of water. This suggests that Triton is made up of a mixture of rock and icy material.

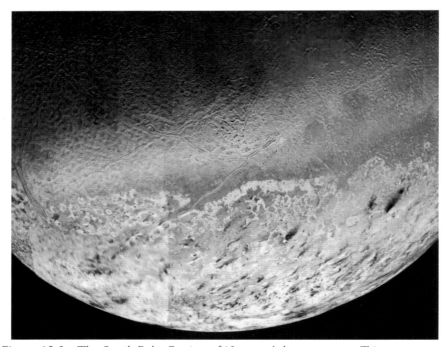

Figure 12.9   The South Polar Region of Neptune's largest moon, Triton. (Photo: NASA)

Triton has a tidal effect on Neptune that is tending to pull Triton towards Neptune. In about a quarter of a billion years, Triton may be pulled apart by Neptune's gravitational pull.

DID YOU KNOW?

Neptune's largest moon, Triton, was discovered in 1846, but it was not until at least the 1930s that it was officially known by that name. Although the name Triton was proposed in 1880 by Camille Flammarion, it was often simply known as 'the satellite of Neptune'. Triton is an odd moon in that its orbit is highly tilted (157°) to the plane of Neptune's equator. It also rotates in a direction opposite to the direction of Neptune's rotation. Triton may have been formed independently of Neptune and later captured by Neptune's gravity. The second of Neptune's moons, Nereid, was discovered in 1949.

Table 12.3 Moons of Neptune

Name of moon	Distance (km)	Period (days)	Diameter (km)	Discovered
Naiad	48 200	0.29	66	1989
Thalassa	50 100	0.31	82	1989
Despina	52 500	0.34	150	1989
Galatea	62 000	0.43	176	1989
Larissa	73 500	0.56	194	1989
Proteus	118 000	1.12	420	1989
Triton	354 800	5.88	2700	1846
Nereid	5 510 000	360	340	1949
Halimede	15 728 000	1880	48	2002
Sao	22 422 000	2914	48	2002
Laomedeia	23 571 000	3167	48	2002
Psamathe	46 695 000	9115	28	2003
Neso	48 387 000	9373	60	2002

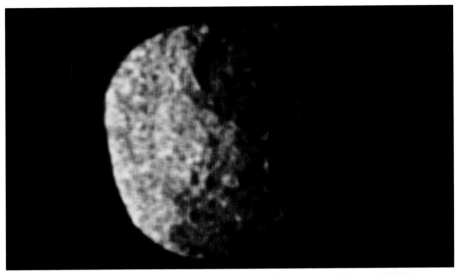

Figure 12.10 Proteus is a dark, cratered moon of Neptune. (Photo: NASA)

The inner four moons, Naiad, Thalassa, Despina and Galatea, orbit within the ring system. Larissa, Proteus, Triton, Nereid and the five small outer moons orbit beyond the ring system. Four of the moons have retrograde motions – Triton, Halimede, Psamathe and Neso. Many of the names of Neptune's moons come from the nereids or water spirits of Greek mythology.

Prior to the visit of Voyager, Nereid was thought to be the second largest moon of Neptune. However, when Voyager discovered Proteus, Proteus was found to be larger than Nereid. Apart from Triton, all the moons of Neptune are irregular in shape, being too small to form uniform spheres.

WEB NOTES

For fact sheets on any of the planets, including Neptune, check out:
    <http://nssdc.gsfc.nasa.gov/planetary/planetfact.html>
    <http://www.space.com/neptune/>
    <http://www.nasm.si.edu/etp/>

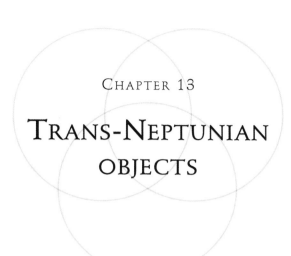

CHAPTER 13

TRANS-NEPTUNIAN OBJECTS

OUR UNDERSTANDING OF the solar system has changed in the last decade, mainly as a result of new discoveries beyond Neptune. This chapter explores these discoveries.

TRANS-NEPTUNIAN OBJECTS

A **trans-Neptunian object** (TNO) is any object in the solar system that orbits the Sun at a greater distance on average than the planet Neptune. A range of objects fall into this category and many have only recently been discovered. By 1 January 2008, astronomers had catalogued over a thousand trans-Neptunian objects.

The first astronomer to suggest the existence of a trans-Neptunian population was Frederick C. Leonard, in 1930. In 1943, Kenneth Edgeworth postulated that, in a region beyond Neptune, the material from the primordinal solar nebula would have been too widely spaced to condense into planets. From this he concluded that the outer region of the solar system should contain a very large number of smaller bodies that from time to time could venture into the inner solar system. In the last few

Figure 13.1   The many recently discovered objects that exist beyond the orbit of Neptune have resulted in the solar system being redefined. The scattered disc is a distant region thinly populated with icy bodies. (Photo: J. Wilkinson)

decades, astronomers have identified three regions that exist beyond Neptune: the **Kuiper belt**, the **scattered disc**, and **Oort cloud**.

Trans-Neptunian objects display a wide range of colours from blue-grey to very red. It is difficult to determine the size of TNOs because they are so far away. For large objects, diameters can be measured precisely by occultation of stars. For smaller objects, diameters need to be estimated by thermal measurements and measurements of relative brightness.

THE KUIPER BELT

The **Kuiper belt** is a vast region of the solar system beyond Neptune's orbit. It is best described as shaped like a flat doughnut around the Sun that extends from 30 AU to 50 AU. It is similar to the asteroid belt, although it is much larger and more massive. Like the asteroid belt, it contains mainly small bodies. Whereas the asteroids are composed mainly of rock and metal, the objects in the Kuiper belt are made mostly of rock and ices such as methane, ammonia and water. The temperature of objects within the belt is around −230°C, so they are very cold.

The belt is named after Dutch-born American astronomer Gerard Kuiper, who first predicted its existence in 1951. The objects in this region are called **Kuiper belt objects** or KBOs. The orbits of many of these objects are highly elliptical and are destabilised by Neptune's gravity.

The most famous KBO is the plutoid Pluto. Apart from Pluto, discovered in 1930, the first KBO was discovered by David Jewitt and Jane Luu in Hawaii in August 1992. This object is called **1992 QB1**. It was found 42 AU from the Sun. Six months later, these two astronomers discovered a second object, **1993 FW**, in the same region.

It is suspected that there may be as many as 35 000 objects in the Kuiper belt with diameters of 100 km or greater, as well as many more smaller objects. Most objects in this belt probably formed at the same distance from the Sun as we find them today.

The Kuiper belt is also thought to be the home of short-period comets (those with periods less than 200 years). **Comets** are small bodies of ice and rock in orbit around the Sun. Many comets have highly elliptical orbits that occasionally bring them close to the Sun. When this happens the Sun's radiation vaporises some of comet's icy material, and a long tail is seen extending from the comet's head and pointing away from the Sun.

The plutoid Pluto and its companion Charon are two of the larger KBOs. Several other large KBOs have been discovered, including **Quaoar** (2002 LM60), **Makemake** (2005 FY9) and **Orcus** (2004 DW).

DID YOU KNOW?

Halley's comet originates in the Kuiper belt and has a period of about 76 years. One of the strange things about Halley's comet is that is motion is retrograde, that is, it moves in the opposite direction to that of the planets. In 1986 five spacecraft were directed towards Halley as it approached Earth. Two of these missions were launched by the USSR (Vegas 1 and 2), two by Japan (Susei and Sakigake), and one by the European Space Agency (Giotto). The probes found that Halley's nucleus was irregular in shape, 16 km long and 8 km wide. The dust-rich surface was darker than coal and contained small hills and craters. The comet's inner coma consists of a mixture of about 80 per cent water vapour, 10 per cent carbon monoxide, and 3.5 per cent carbon dioxide, with some complex organic compounds. Most of the particles in the tail are composed of a mixture of hydrogen, carbon, nitrogen, oxygen and silicates.

Figure 13.2   Halley's comet as seen from Australia in 1986. (Photo: J. Wilkinson)

PLUTO

Pluto was once classified as the ninth major planet of the solar system but was reclassified in 2006 by the IAU as a dwarf planet before the new category of 'plutoid' was established in 2008. The main reason why Pluto was demoted from a major planet was that it has not cleared the neighbourhood around its orbit. Pluto orbits the Sun in the inner Kuiper belt in a region where many other objects also orbit.

Pluto is much smaller than any of the official planets, and is even smaller than seven of the moons in the solar system. It is so small and distant that we cannot see any surface detail on the planet through Earth-based telescopes. It has a diameter of 2320 km and takes 248 years to travel once around the Sun. The orbit of this body is on a different plane from those of the true planets.

Pluto orbits the Sun at an average distance of about 5913 million kilometres, making it about 40 times more distant than Earth from the

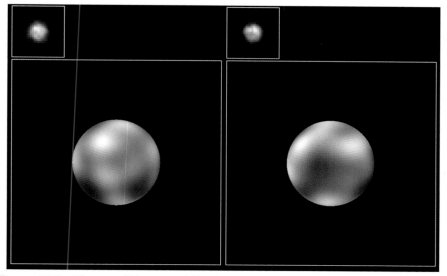

Figure 13.3   Photos of Pluto are difficult to obtain because it is so far away and no space probe has yet visited it. This image has been made up from a mosaic of photos. (Photo: NASA)

Sun. It is so distant that Pluto's brightest daylight is less than moonlight on Earth. Pluto is always further from the Sun than is Uranus, but every 248 years it moves inside Neptune's orbit for about a 20-year period, during which time Pluto is closer to the Sun than is Neptune. Pluto crossed Neptune's orbit on 23 January 1979 and remained within it until 11 February 1999.

Early views about Pluto

In Roman mythology Pluto (Greek: Hades) was the god of the underworld. The planet received this name in part because it was so far from the Sun and was in perpetual darkness.

Early astronomers did not know about Pluto because it could not be seen from Earth by the unaided eye. It is even difficult to locate using Earth-based telescopes.

The discovery of Pluto is an interesting story. Irregularities in the orbits of Uranus and Neptune led to the suggestion by US astronomers Percival Lowell and William Pickering that there might be another body (planet X) orbiting beyond Neptune. Lowell died in 1916, but he initiated the construction of a special wide-field camera to search for planet X. In 1930, Clyde W. Tombaugh at Lowell Observatory in Arizona found planet X, which was later named Pluto. As it turned out, Pluto was too small and too distant to influence the orbits of Uranus and Neptune, and the search for another planet continued. The name 'Pluto' also honours Percival Lowell, whose initials PL are the first two letters of the name.

At one time it was thought that Pluto may have once been a moon of Neptune, but this now seems unlikely. A more popular idea is that Triton was once a Kuiper belt object like Pluto, moving in an independent orbit around the Sun, and was later captured by Neptune.

Table 13.1 Details of Pluto

Distance from Sun	5 913 520 000 km (39.5 AU)
Diameter	2320 km
Mass	1.27×10^{22} kg (0.002 times Earth's mass)
Density	2.03 g/cm³ or 2030 kg/m³
Orbital eccentricity	0.248
Period of revolution (length of year)	90 740 Earth days or 248 Earth years
Rotation period	6.38 Earth days
Orbital velocity	17 064 km/h
Tilt of axis	122.5°
Average temperature	−220°C
Number of moons	3
Atmosphere	nitrogen, methane
Strength of gravity	0.67 N/kg at surface

Probing Pluto

As of 2007, Pluto has not yet been visited by a space probe from Earth. Even the Hubble Space Telescope can resolve only the largest features on the surface of Pluto. A space probe called **New Horizons** was launched in January 2006 on a mission to Pluto and beyond. To get to Pluto, this spacecraft will get a gravity assist from Jupiter. New Horizons will pass through the Jupiter system at 82 000 km/h, ending up on a path to reach Pluto in 2015 (see Figure 13.3).

Position and orbit

Pluto orbits the Sun in an elliptical path at an average distance from the Sun of about 5913 million kilometres. Because it is so far from the Sun, Pluto takes a very long 248 years to go around the Sun once. Its orbit is

Table 13.2 Significant space probes to Pluto

Probe	Country of origin	Launched	Comments
New Horizons	USA	2006	First space probe to fly by Pluto, expected in 2015

inclined at an angle of 17.2° to the orbital plane of the true planets. This means that its orbit rises above and drops below the ecliptic plane. Pluto's orbit is so elliptical that for 20 years of its orbital period it is closer to the Sun than is Neptune, which follows a near-circular orbit.

Like Uranus, Pluto is also tipped over on its side. Its rotational axis is inclined at an angle of 122.5° to the plane of its orbit. This means that Pluto's equator is almost at right angles to the plane of its orbit. Pluto also rotates in the opposite direction to most of the other planets, with one rotation taking 6 days, 9 hours and 18 minutes.

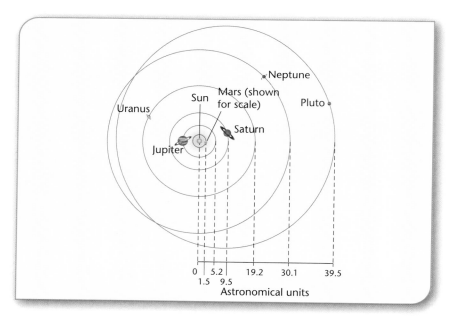

Figure 13.4 Orbital path of Pluto. Notice how different this is from the paths of the major planets.

DID YOU KNOW?

At its closest approach to the Sun, Pluto is 30 times more distant from the Sun than is Earth. At its greatest distance from the Sun, Pluto is 50 times more distant from the Sun than is Earth. Pluto will next be at its maximum distance from the Sun during the year 2113. During the coldest 124 years of its orbit, all of Pluto's atmosphere condenses and falls to the surface as frost.

Images taken of Pluto by the Hubble Space Telescope have shown that the reflectivity of its surface varies. Lighter areas are probably patches of nitrogen and methane frost as well as exposed regions of water ice.

Density and composition

The high average density of Pluto (2.03 g/cm^3) indicates its composition is a mixture of about 70 per cent rock and 30 per cent water ice, much like Triton. One theory is that Pluto and Triton formed at the same time in the same part of the solar nebula.

Pluto probably has a large rocky core of silicate materials, surrounded by a mantle rich in ices and frozen water. The extent of Pluto's crust is

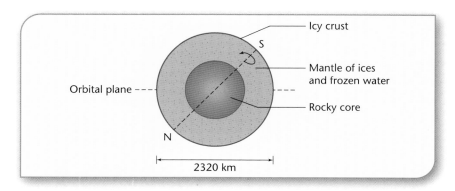

Figure 13.5   Internal structure of Pluto.

unknown, but it is thought to be covered in patches of frozen nitrogen, water, methane and ethane. Pluto's spectrum shows absorption lines of various solid ices that cover the planet's surface, including nitrogen, methane and carbon monoxide.

The strength of gravity on Pluto is much less than Earth's gravity (0.67 N/kg compared to 9.8 N/kg). This means that a 75 kg person weighing 735 N on Earth would weigh only 50 N on Pluto.

The surface

Astronomers know little about the surface of Pluto because the body is so far from Earth. Pluto's diameter is less than one-fifth that of Earth so it is difficult to observe anything on a surface that is so far away. The best views of Pluto show a brownish disc with bright and dark areas (Figure 13.3). The bright areas are probably covered with frozen nitrogen, with smaller amounts of methane, ethane and carbon monoxide. The composition of the darker areas on the surface is unknown but they may be caused by decaying methane or other carbon-rich material.

Pluto's interior is probably rich in ices with some frozen water. The central core probably contains solid iron and nickel and rocky silicate materials.

The atmosphere

Little is known about Pluto's atmosphere but it thought to contain about 98 per cent nitrogen with about 2 per cent methane and carbon monoxide. The composition was determined from observations made when the planet passed in front of a bright star (an occultation).

The atmosphere is tenuous and the pressure at the surface is only a few millionths of that of Earth. It is thought to extend about 600 kilometres above the surface. Because of Pluto's elliptical orbit, the atmosphere may be gaseous when Pluto is nearest the Sun and frozen when furthest from the Sun. NASA wants its New Horizon space probe to arrive at Pluto when its atmosphere is unfrozen.

Pluto's weak gravity means its atmosphere extends to a greater altitude than does Earth's atmosphere.

Temperature and seasons

The surface temperature on Pluto varies between about −235°C and −210°C. The warmer regions roughly correspond to the darker regions on the surface.

Because the orbit of Pluto is so elliptical, the amount of solar radiation it receives varies markedly between its extreme positions. Its 248-year orbital period means that any seasonal change is very slow to take place.

Magnetic field

Pluto may have a magnetic field, but it would not be strong. The New Horizons probe may be able to take measurements of any field it may have.

Moons of Pluto

In 1978 the American astronomer James Christy noticed that Pluto had an elongated shape in photographs. A search through previous images also showed a similar shape. This observation led to the discovery of a moon orbiting Pluto. The moon was named **Charon**, and it was found to orbit Pluto at a distance of 19 700 km over a period of 6.39 days. It turned out that Pluto and Charon rotate synchronously, each keeping the same face to the other. Astronomers were able to observe the two bodies rotating around each other during 1985 and 1990 when the two were edge-on to Earth. Such observations enabled astronomers to determine that Pluto's diameter was 2320 km and Charon's 1270 km. Because these sizes are close, some astronomers referred to the two bodies as a double planet.

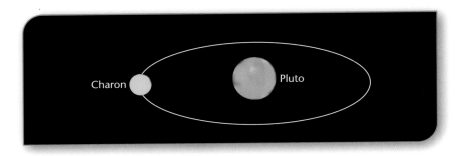

Figure 13.6 Charon in orbit around Pluto. Charon rotates on its axis in the same time as it takes to orbit Pluto. Charon therefore shows the same face towards Pluto and would remain fixed at the same point in the sky.

The average distance between Charon and Pluto is one-twentieth the distance between the Earth and our Moon. The combined masses of Pluto and Charon amount to less than one four-hundredth of Earth's mass.

The best pictures of Pluto and Charon have come from the Hubble Space Telescope. Both bodies are thought to consist of rock and ice. Charon's surface is probably covered with dirty water ice, which is why it doesn't reflect as much light as does Pluto. The Hubble observations show that Charon is bluer than Pluto.

DID YOU KNOW?

The New Horizons spacecraft is expected to start taking data on Pluto and Charon months before the spacecraft arrives at Pluto. On the way, the spacecraft will look for ultraviolet emissions from Pluto's atmosphere and make global maps of Pluto and Charon in green, blue, red and a wavelength that is sensitive to methane frost. The probe will also take spectral maps in the near-infrared spectrum to help scientists determine the surface composition and temperatures. The probe will approach within 9600 km of Pluto and about 27 000 km of Charon.

Table 13.3 The moons of Pluto

Name	Distance from Pluto (km)	Period (days)	Diameter (km)	Discovered
Charon	19 700	6.39	1270	1978
Nix	49 500	25.5	60–200	2005
Hydra	64 500	38.0	60–200	2005

In late 2005, a team of scientists using the Hubble Space Telescope discovered two additional moons orbiting Pluto. Provisionally named S/2005 P1 and S/2005 P2, they are now known as **Nix** and **Hydra**. They are estimated to be between 60 km and 200 km in diameter. These two tiny moons are roughly 5000 times fainter than Pluto and between 49 000 km and 65 000 km away from Pluto.

OTHER KUIPER BELT OBJECTS

Since the 1980s, hundreds of icy bodies have been detected in the Kuiper belt. Most of these objects are much smaller than Pluto. Once the orbit of a KBO is determined, it is given an official number by the Minor Planet Centre of the IAU. As of early 2008, at least 84 objects have numbers and many of those have names, including: Asbolus, Bienor, Chaos, Chariklo,

Figure 13.7   Pluto (centre of diffraction lines) and its three moons, Charon, Nix and Hydra. (Photo: NASA)

Table 13.4 Main Kuiper belt objects (as of 2008)

Name	Diameter (km)	Orbital radius (AU)	Orbital period (years)	Orbital eccentricity	Number of moons
Pluto	2320	39.5	248	0.249	3
Ixion	800	30–49	250	0.242	0
Varuna	900	43	283	0.051	0
Quaoar	1250	42	288	0.034	1
2002 AW197	750	41–53	325	0.132	0
2003 EL61	1500	40–90	285	0.189	2
Orcus	950	30–48	247	0.225	1
Makemake	1500	38–53	308	0.159	0

Chiron, Cyllarus, Deucalion, Elatus, Huya, Hylonome, Ixion, Nessus, Okyrhoe, Orcus, Pelion, Pholus, Quaoar, Phadamanthus, Thereus and Varuna. Details are known of only a few of the KBOs.

Ixion (2001 KX76) was discovered on 22 May 2001. Its estimated diameter is 800 km and its distance from the Sun varies between 30 and 49 AU because of a highly elliptical orbit. It has a reddish colour and spectroscopic data suggests its composition is a mixture of dark carbon and tholin (a tar-like substance formed by irradiation of water and carbon-based compounds). Ixion and Pluto follow similar but differently orientated orbits. Ixion's orbit is below the ecliptic, whereas Pluto's is above it. Ixion takes about 250 years to orbit the Sun.

Varuna (2000 WR106) is a small KBO named after the Hindu god of the sky, rain, oceans and rivers; law and the underworld. Varuna was discovered in November 2000 by R. McMillan and has a size of about 900 km. It orbits the Sun at an average distance of 43 AU in a near-circular orbit. Unlike Pluto, which is in 2 : 3 orbital resonance with Neptune, Varuna is free from any significant perturbation from Neptune.

Varuna has a fast rotational period of about 6 hours. The surface of this body is red, and it is dark compared with other KBOs, suggesting the surface is largely devoid of ice.

One of the largest KBOs is **Quaoar** (2002 LM60), discovered in 2002 by astronomers Chad Trujillo and Mike Brown in California, USA, using large ground-based telescopes. The name Quaoar is derived from the Native American Tongva people, from the area around Los Angeles where the discovery was made. Quaoar is reddish in colour and orbits the Sun once every 288 years in a near-circular orbit of radius about 42 AU. It has an estimated diameter of about 1250 km, roughly the size of Pluto's moon Charon and about one-tenth the size of Earth. Although smaller than Pluto, Quaoar is 100 million times greater in volume than all the asteroids combined. Quaoar is spherical and is a possible candidate for classification as a plutoid. Like other KBOs, Quaoar's composition is thought to be mainly ice mixed with rock. The surface temperature is estimated at −230°C, making it one of the coldest bodies in the solar system. From the surface of this body, the distant Sun would appear only as bright as Venus does from Earth. A satellite of about 100 km diameter was discovered orbiting Quaoar in February 2007, but little is known about it.

Figure 13.8 Artist's impression of what Quaoar, one of the largest KBOs, might look like. (Photo: NASA)

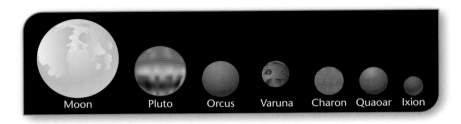

Figure 13.9   Comparison of the sizes of major Kuiper belt objects with Earth's Moon.

Another KBO is **2002 AW197**, which was discovered in January 2002 by a group of scientists led by Mike Brown. This object is about 750 km in diameter and orbits on an elliptical path between 41 and 53 AU from the Sun, taking about 325 years for one orbit. Spectroscopic analysis shows a strong red colour but no water ice.

One of the strangest objects in the Kuiper belt is **2003 EL61**. This object is not quite as large as Pluto and it is oval-shaped like an Australian or American football. It spins end-over-end every four hours like a football that has been kicked. 2003 EL61 appears to be made almost entirely of rock, but with a glaze of ice over its surface. In 2005 astronomers detected a moon orbiting 2003 EL61. By following the orbit of the moon, astronomers have been able to determine that the mass of 2003 EL61 is about 32 per cent that of Pluto. More recently, a second moon, smaller and fainter than the first, has also been detected.

The odd shape of 2003 EL61 is thought to have been caused by a collision with another object early in its history. This collision knocked some of the original ice away from the surface of the body, leaving behind mostly rock to take the shape we see today. The impact also caused the body to spin rapidly. The small moons orbiting 2003 EL61 may have come from debris blown away during the collision.

In March 2007, astronomers announced they had found five other icy bodies in orbits similar to 2003 EL61. Such families of objects are common in the asteroid belt, but this is the first group found in the Kuiper belt. All

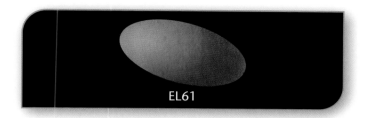

Figure 13.10   The object 2003 EL61 (now Haumea) is shaped like a football.

the fragments have a colour and proportion of water to ice similar to 2003 EL61, and each fragment also has a surface that looks like it was once an internal portion of the original object.

Orcus (2004 DW) is a KBO discovered by Mike Brown and David Rabinowitz (USA) in February 2004. This body has an elliptical orbit around the Sun, similar to that of Pluto, and it takes about 247 years to orbit the Sun. At its closest approach it is 30 AU from the Sun, while its greatest distance is 48 AU. With a diameter of about 950 km, Orcus is smaller than Quaoar. Its surface temperature is around −230°C. Observations in infrared by the European Southern Observatory give results consistent with mixtures of water ice and carbon-based compounds. Orcus appears to have a neutral colour in comparison with the reds of other KBOs. In February 2007, a satellite of size about 220 km was discovered orbiting Orcus. Under the guidelines of the IAU naming conventions, objects with a similar size and orbit to that of Pluto are named after underworld deities; Orcus is a god of the dead in Roman mythology.

The KBO **2005 FY9** (now Makemake) is a large spherical object with a diameter of about 1500 km. Discovered in 2005 by the team led by Mike Brown, this object orbits the Sun once every 308 years in an eccentric and inclined orbit like Pluto's. Spectral analysis showed the surface resembles that of Pluto but is redder. The infrared spectrum indicates the presence of methane, as observed on both Pluto and Eris. No satellites have yet been detected around Makemake. In July 2008 Makemake was classified as a plutoid.

DID YOU KNOW?

Some planetary scientists think that Phoebe, the outermost satellite of Saturn, was probably a typical Kuiper belt object before gravitational perturbations pushed it closer to the Sun and a later interaction left it loosely bound to Saturn. Images of Phoebe taken by the Cassini space probe show its surface to be heavily cratered like our Moon. White material visible on the cliff slopes and cratered walls is probably relatively fresh ice. Phoebe is about 220 km wide.

THE SCATTERED DISC

The **scattered disc** is a sparsely populated region beyond the Kuiper belt, extending from 50 AU to as far as 100 AU and further. Objects in this region have highly eccentric orbits and are often wildly inclined to the orbital plane of the major planets. Two of the first scattered disc objects (SDO) to be recognised are 1995 TL8 (at 53 AU from the Sun) and 1996 TL66 (at 83 AU). Other objects detected include 1999 TD10, 2002 XU93 and 2004 XR190 (at 58 AU). Some astronomers prefer to use the term 'scattered Kuiper belt objects' for objects in this region.

Many of the SDOs are doomed in the long term because, sooner or later, their highly eccentric orbits will carry them close to the giant planets to undergo more scattering. They may last a few million years or even 100 million years in their current orbits, but eventually Neptune will flip them nearer Uranus, Saturn or Jupiter. These planets will, in turn, fling them outward, far beyond the Kuiper belt and into the Oort cloud or out of the solar system entirely, or closer to the Sun (where they will become comets).

One of the major scattered disc objects is **Eris** (2003 UB313, previously known as Xena). The elliptical orbit of this body takes it to within 36 AU from the Sun and as far out as 100 AU. It was the discovery of this object in 2003 that prompted astronomers to refine the definition of a planet. If

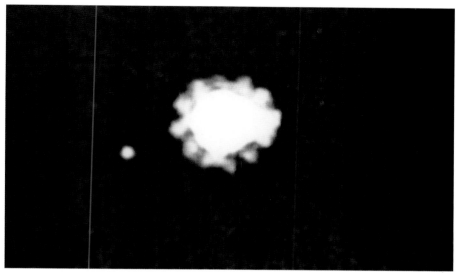

Figure 13.11   Eris, the largest known scattered disc object, and its moon Dysnomia. (Photo: NASA)

Eris had been classed as a planet, there may have been as many as fifteen planets in the solar system. In the end, the IAU decided on a definition that excluded Eris and also Pluto as major planets, instead classifying them as plutoids.

Eris has a diameter of 2400 km, which makes it larger than Pluto. It is the largest object found in orbit around the Sun since the discovery of Neptune and its moon Triton in 1846. At times Eris is even more distant than the Oort cloud object Sedna (see below), and it takes more than twice as long to orbit the Sun as Pluto (560 years). In 2005, a near-infrared spectrograph on the Gemini Telescope in Hawaii showed the surface of Eris to be mainly methane ice. Methane ice suggests a primitive surface unheated by the Sun since the solar system formed. If Eris ever had been close to the Sun, the methane ice would have boiled off. The interior of the plutoid is probably a mix of rock and ice, like Pluto. The elliptical orbit of Eris is tilted at an angle of 45° to the orbital plane of the major planets. Eris has also been found to have a moon, named **Dysnomia**.

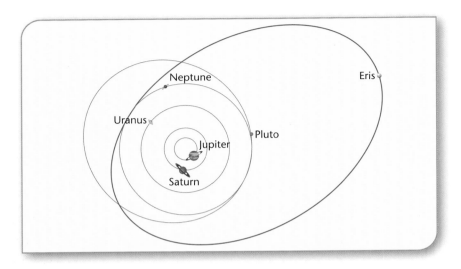

Figure 13.12   Orbital path of Eris (2003 UB313).

Eris is currently about three times Pluto's distance from the Sun, following an orbit that is about twice as eccentric and twice as steeply inclined to the plane of the solar system.

The Oort cloud

The **Oort cloud** is an immense spherical shell surrounding the solar system between 1000 and 100 000 AU (30 trillion km) from the Sun. This region contains billions of small icy objects probably left over from the formation of the solar system. Sometimes the orbit of one of these objects gets disturbed by other bodies, causing it to come streaking into the inner solar system as a long-period comet (one with a period of around 2000 years). In contrast, short-period comets take less than 200 years to orbit the Sun and they come from the Kuiper belt. The total mass of comets in the Oort cloud is estimated to be 40 times that of Earth.

One of the major Oort cloud objects is **Sedna** (2003 VB12), which was discovered in November 2003 by a team led by Mike Brown at Palomar

Observatory near San Diego, California, USA. The object was named after Sedna, the Inuit goddess of the sea, who was believed to live in the cold depths of the Arctic Ocean. Sedna has a highly elliptical orbit that is inclined at about 12° to the ecliptic. Its distance from the Sun varies between 76 AU and 975 AU, so it is best described as an inner Oort Cloud Object. Sedna will make its closest approach to the Sun (perihelion) about the year 2076, and will be furthest from the Sun (aphelion) in 8207. The shape of its orbit suggests it may have been captured by the Sun from another star passing by our solar system, or its orbit could be affected by another larger object further away in the Oort cloud.

Sedna is an odd body because astronomers thought they would not find an object like it in the empty space between the Kuiper belt and the Oort cloud. Sedna is the second reddest object in the solar system after Mars. Its size is estimated to be between 1200 and 1800 km (about three-quarters the size of Pluto), and it takes 12 000 years to orbit the Sun. Recent estimates put its rotational period at about 10 hours and its surface temperature at a cold –250°C. Sedna appears to have very little methane ice or water ice on its surface.

THE FUTURE

There is no doubt that more discoveries will be made in the Kuiper belt and the Oort cloud. Any new objects discovered will of necessity be faint,

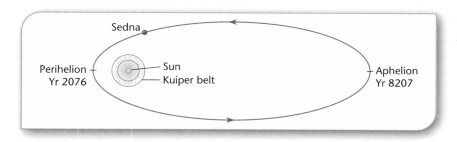

Figure 13.13   Orbital path of Sedna.

Table 13.5 Plutoids (as at 2008)

Name	Diameter (km)	Orbital radius (AU)	Orbital period (years)	Orbital eccentricity	Number of moons	Location
Pluto	2320	39.5	248	0.249	3	Kuiper belt
Eris	2400	36–100	560	0.442	1	Scattered disc
Makemake	1500	38–53	308	0.159	0	Kuiper belt

cold, and far away from Earth. It is also possible that the IAU will change their definition of what constitutes a planet in the future.

In August 2006, the IAU reclassified a number of objects in the solar system as 'dwarf planets', but then decided on a new category of 'plutoid' for the larger TNOs. Currently three plutoids are recognised by the IAU: Pluto, Makemake and Eris. Several other objects in the Kuiper belt are under consideration, with as many as 50 that could eventually qualify. Plutoids and dwarf planets share a number of characteristics with normal planets, but they are not dominant in their orbit around the Sun. The plutoids and dwarf planets classified so far are members of larger populations. For example, the dwarf planet Ceres is the largest body in the asteroid belt, the plutoid Pluto is the largest body in the Kuiper belt, and the plutoid Eris is a member of the scattered disc.

Under the 2008 IAU definition of planet, there are currently eight planets, one dwarf planet, and three plutoids in the solar system. There has been some criticism of the new definition, and some astronomers have even stated that they will not use it. Part of the dispute centres around the idea that plutoids and dwarf planets should be classified as normal planets. For now, the reclassification of Ceres, Pluto and Eris has attracted much media and public attention.

WEB NOTES

For further information on objects in the solar system beyond Neptune, try:
<http://solarsystem.nasa.gov/>
Click on 'planets' and then use the menu to select object.
For up-to-date information on space news, including trans-Neptunian objects, try:
<http://www.spacetoday.org>
<http://www.space.com/planets/>

CHAPTER 14

OVERVIEW AND THE FUTURE

THIS CHAPTER PROVIDES an overview of our current knowledge of the solar system and the ways we classify objects within it. Motivated by the recent discoveries in our own solar system and a desire to search for extraterrestrial life, astronomers are also searching other stars for planetary systems. The characterisation of planetary systems outside our own is an emerging field of study that will draw much attention and interest in the future.

TODAY'S SOLAR SYSTEM

The solar system consists of the Sun, eight major planets, at least three dwarf (or minor) planets, and more than 170 moons or natural satellites. A large number of small bodies exist in the asteroid belt, Kuiper belt and scattered disc. The whole solar system is surrounded by the Oort cloud. It seems as though the solar system is fairly full, but most of the solar system is actually empty space. The planets are very small in comparison to the space between them.

In the previous chapters we have seen that new discoveries have resulted from sending space probes to explore the various planets and moons of the solar system. Most discoveries have concerned the planets Venus, Mars,

Figure 14.1 Our solar system is part of the Milky Way galaxy. This photograph shows a section of the Milky Way as we see it from Earth at night. There are probably countless suns and countless 'earths' all rotating around their suns in exactly the same way as the eight planets of our system. (Photo: J. Wilkinson)

Jupiter and Saturn. It takes time to send space probes to the more outer planets, and so we know less about them. In the next couple of decades we will learn more about the outer planets, and our understanding will improve as technology improves.

The orbits of the major planets in our solar system are described as ellipses with the Sun at one focus, though all orbits except that of Mercury are very nearly circular. The orbits of the major planets are all more or less in the same plane (called the ecliptic). The ecliptic is inclined only 7° from the plane of the Sun's equator. The planets all orbit in the same direction (anti-clockwise looking down from above the Sun's north pole), and all the major planets apart from Venus and Uranus rotate in that same direction.

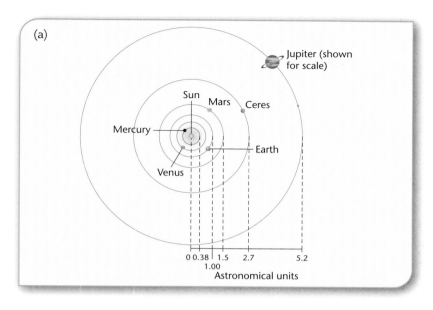

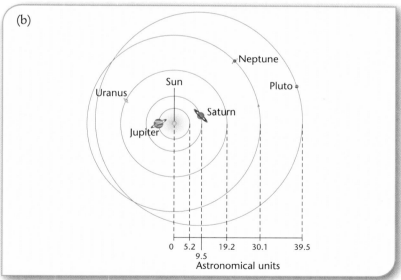

Figure 14.2 Orbits of the planets in (a) the inner solar system and (b) the outer solar system. The dwarf planet Ceres and plutoid Pluto have been included for comparison of orbits.

A large number of asteroids orbit the Sun in two regions, the asteroid belt (which lies between the orbits of Mars and Jupiter) and the Kuiper belt (which lies beyond Neptune).

The inner asteroids are rocky bodies, while the outer ones in the Kuiper belt are icy and rocky. Bodies in these two regions are probably the remains of material that failed to condense into a spherical shape to form a planet; that is, they are left-over material from the formation of the solar system.

Short-period comets appear to originate from the Kuiper belt, while long-period comets come from the Oort cloud. Comets are cold icy objects with elongated orbits. They sometimes pass into the inner solar system, and when they are close to the Sun we can see the trail they leave.

The classification of objects in the solar system has changed in recent years as new objects have been discovered. From 1930 until recently there were nine objects classed as planets, but after a change of definition by the IAU in 2006 and recent discoveries of objects beyond Neptune, there are now eight major planets, one dwarf planet and three plutoids. It is possible that the various objects will be reclassified in the future as further discoveries are made, but this is the way science evolves.

Various methods have been used to classify the objects in the solar system. The eight major planets can be classified by position from the Sun, size, composition, and by history and type.

By position from the Sun

- inner planets: Mercury, Venus, Earth and Mars
- outer planets: Jupiter, Saturn, Uranus and Neptune

The asteroid belt separates the inner and outer planets.

By size

- small planets: Mercury, Venus, Earth and Mars
- giant planets (gas giants): Jupiter, Saturn, Uranus and Neptune

By composition

- terrestrial or rocky planets: Mercury, Venus, Earth and Mars

The terrestrial planets are composed primarily of rock and metal and have high densities, slow rotation, solid surfaces, no rings and few moons.

- Giant gas planets: Jupiter, Saturn, Uranus and Neptune

The giant gas planets are composed primarily of hydrogen and helium and generally have low densities, rapid rotation, deep atmospheres, rings and lots of moons.

By history

- classical planets: Mercury, Venus, Mars, Jupiter and Saturn

These have been known since ancient times and are visible to the unaided eye from Earth.

- modern planets: Uranus and Neptune

These have been discovered in modern times and are visible from Earth only with the aid of telescopes.

By current type

- major planets: Mercury, Venus, Earth, Mars, Jupiter, Saturn, Uranus and Neptune
- dwarf planets: Ceres
- plutoids: Pluto, Eris and Makemake
- smaller bodies: asteroids, moons

ORDER AND DISORDER

The most obvious feature of the solar system is that the planets constitute an orderly system. All of them move about the Sun in the same direction in nearly circular orbits and in nearly the same plane. The spacing between them increases with distance in an orderly way, and most planets rotate in the same direction as they revolve.

While the system of main planets and their satellites is characterised by order, that of the comets and meteoroids, and objects in the Kuiper belt, scattered disc and Oort cloud are often characterised by disorder – these objects move in elongated, tilted orbits.

Astronomers are attempting to understand this order and disorder by sending space probes to explore the solar system further. Unravelling the origin and history of the solar system provides one of the great adventures of science.

BEYOND THE SOLAR SYSTEM – EXOPLANETS

Our Sun and its system of planetary bodies are not alone in the universe. Since 1995, astronomers have discovered more than 270 planets in orbit

Figure 14.3   The Pleiades (or Seven Sisters) is a cluster of young stars formed at the same time from a gas cloud of uniform composition. It lies about 400 light years from Earth in the constellation Taurus. The blue haze around the stars, called a reflection nebula, is caused by interstellar dust that scatters and reflects blue light. There may be planetary systems forming around some stars in this cluster. (Photo: J. Wilkinson)

around other stars. Most of the stars with planets are similar in size to our Sun. Several of these stars have more than one or two planets orbiting them. Since our Sun is a fairly typical star in the Milky Way galaxy, one would expect that other planetary systems would be common. Planets that exist around other stars are called **extra-solar planets** or **exoplanets**. Part of the enormous attraction of searching for exoplanets is the possibility that some of these new-found worlds will be able to support life.

Detecting exoplanets

Astronomers detect planets around other stars by a number of methods. One method involves accurately measuring changes in the intensity of light coming from a star as a planet passes in front of the star (the transit technique). If such changes are regular, then it is likely a planet is orbiting the star. Another method involves an accurate analysis of the motion of the star – a wobble in a star's motion could be caused by the gravitational pull of its planets. Astronomers also use spectral analysis to detect exoplanets. An exoplanet's properties are determined by combining information about a planet's brightness, colour, spectral properties and the variability of these over time.

Detection of planets around a distant star is extremely difficult from Earth because the intense brightness of a star tends to block out any planets. Very high resolution instrumentation is needed to separate the planet from its parent. Most of the known exoplanets have been discovered using Earth-based telescopes, but astronomers also use space telescopes (that is, those in orbit above Earth's atmosphere).

Most of the exoplanets astronomers have already discovered have been around stars similar to our Sun. It is estimated that at least 12 per cent of sun-like stars have planets orbiting them. Most known exoplanets are also 'giant' planets, typically as large or larger than Jupiter. The difficulty of detecting exoplanets means that current techniques are suited to discovering large exoplanets very close to their parent stars.

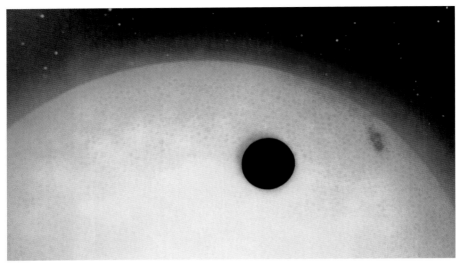

Figure 14.4 The star HD189733, an orange dwarf 63 light years away in Vulpecula, has a Jupiter-mass planet very closely circling it every 2.2 days and transiting the star's face once per orbit (artist's interpretation). (Photo: J. Wilkinson)

Probing for exoplanets

Space telescopes orbiting Earth are used to search for exoplanets. Examples include the Hubble Space Telescope (launched in 1990), Spitzer Space Telescope (2003), MOST (Microvariability and Oscillation of STars, 2003), and the French and European Space Agency's COROT (COnvection, ROtation and planetary Transits, 2006).

NASA's Kepler Mission and Space Interferometry Mission (SIM) and the ESA's Eddington Mission are designed to search for terrestrial planets in the Sun's neighbourhood of the Milky Way galaxy. These probes are scheduled to be launched before 2010 and are the first missions capable of finding Earth-size and smaller planets around other stars via space telescopes. The **James Webb Space Telescope** is scheduled for launch by NASA and the ESA in 2013.

NASA's **Terrestrial Planet Finder** and the European Space Agency's **Darwin** are two other missions designed to search for exoplanets. These

Table 14.1 Significant missions to search for exoplanets

Mission	Country of origin	Date of origin	Comment
Hubble Space Telescope	USA	1990	In orbit around Earth
Spitzer Space Telescope	USA	2003	In Earth orbit
MOST	Canada	2003	
COROT	France, ESA	2007	
Kepler	USA	2009[a]	One-metre space telescope
SIM	USA	2010–15[a]	
Eddington	ESA	2010[a]	Space telescope to detect light changes in stars
James Webb	USA, ESA, Canada	2013[a]	Infrared space telescope
Terrestrial Planet Finder	USA	2015[a]	
Darwin	ESA	2015[a]	Four or five free-flying space telescopes

a   Future mission

DID YOU KNOW?

Since its launch in 2003, the **Spitzer Space Telescope** has become a major tool in the quest to characterise exoplanets and the protoplanetary disks where planets are formed. Spitzer obtains images and spectra by detecting the infrared energy radiated by objects in space between wavelengths of 3 and 180 microns (1 micron is one-millionth of a metre). Most infrared radiation of this wavelength is blocked by Earth's atmosphere and so cannot be observed from the ground. Consisting of a 0.85-metre telescope and three cryogenically cooled scientific instruments, Spitzer is the largest infrared telescope ever launched into space. Its highly sensitive instruments provide a unique view of the universe and allow astronomers to peer into regions of space that are hidden from optical telescopes. The mission is the fourth and final observatory under NASA's Great Observatories program, which also includes the Hubble Space Telescope, the Chandra X-Ray Observatory and the Compton Gamma Ray Observatory. Spitzer was able to detect water vapour on an alien world, and created the first exoplanet weather map.

programs are planned for about 2015. The Darwin mission consists of four or five free-flying spacecraft and will use the infrared part of the spectrum to examine exoplanets for the same gases present in Earth's atmosphere. It is estimated that we will find a twin planet to Earth at some time in the next decade. In this time many new planetary systems will be discovered. Some of these will undoubtedly be similar to our own solar system.

Notable exoplanet discoveries

In 1988 the Canadian astronomers Bruce Campbell, Gordon Walker and Stevenson Yang made the first discovery of an exoplanet. Their radial velocity observations suggested that a planet orbited the star Gamma Cephei. In 2003, improved techniques allowed the discovery to be confirmed.

In early 1992, radio astronomers Aleksander Wolszczan and Dale Frail announced the discovery of planets around the **pulsar PSR1257+12**. This discovery was quickly confirmed, and is generally considered to be the first definitive detection of an exoplanet. Pulsar planets are thought to have formed from the remains of the supernova that formed the pulsar.

The first long-period exoplanet to be discovered was Ursae Majoris b in 1996, orbiting in a near circular orbit at 2.11 AU from its parent star.

In 1998, **Gliese 876b** became the first exoplanet discovered orbiting a red dwarf star (Gliese 876). The first multiple planetary system was detected in 1996 around the star Upsilon Andromedae. It contains three planets, all Jupiter-like. Planets b, c and d were announced in 1996, 1999 and 1999 respectively. (Lower-case letters assigned to exoplanets are based on the order in which they were discovered, and not on their position.)

In December 2000, astronomers using the 3.9-metre Anglo-Australian Telescope at Siding Spring discovered a planet around the star **Mu Arae**. Since then, three more planets have been discovered around this star. Mu Arae is a sun-like star 50 light years away in the constellation Ara. Its planetary system consists of:

- a Uranus-mass planet orbiting extremely close to the star (0.1 AU and 9-day period)
- a 0.5 Jupiter-mass planet in an almost Earth-like orbit (0.9 AU and 310-day period)
- a 1.7 Jupiter-mass planet at 1.5 AU with a period of 643 days
- a 1.8 Jupiter-mass planet in a Jupiter-like orbit

In 2001, astronomers using the Hubble Space Telescope announced that they had detected an atmosphere around the exoplanet **HD209458 b**. Using spectral analysis, they detected sodium in the atmosphere. In the same year the first planet was discovered around a giant orange star, Iota Draconis. The planet is very massive and has a very eccentric orbit.

Astronomers at the European Southern Observatory made the first confirmed image of an extrasolar planet in 2005. The planet is orbiting a brown dwarf star named **2M1207**. The planet is five times the size of Jupiter and almost twice as far from its star as Neptune is from our Sun.

Figure 14.5   Research at the Anglo-Australian Optical Telescope at Siding Spring, Australia, includes the search for exoplanets. (Photo: J. Wilkinson)

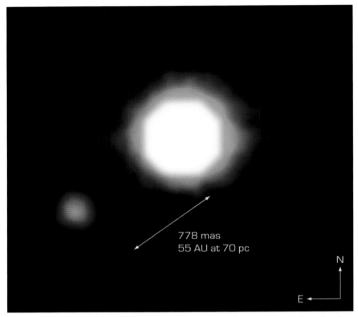

778 mas
55 AU at 70 pc

N

E

Figure 14.6 The first confirmed image of an exoplanet, taken by the European Space Observatory in Chile. The star is a brown dwarf 2M1207 and is shown as blue in this infrared photograph. The exoplanet is 55 AU from its star. (Photo: European Organisation for Astronomical Research in the Southern Hemisphere)

Between the years 1996 and 2006, astronomers found that the star **55 Cancri** has five planets. This star is a yellow dwarf with a red companion. Three of the planets of 55 Cancri orbit between 0.04 and 0.24 AU, another orbits at 0.8 AU, and the outermost one orbits at a Jupiter-like distance of 5.8 AU.

In 2006, the first exoplanet to orbit a bright familiar star was discovered by two independent groups of astronomers in Germany. The star is Pollux, in the constellation Gemini. The planet has a mass of about three Jupiters and orbits Pollux in a near-circular orbit with a 590-day period. This star can be seen with the unaided eye from Earth but its planet cannot be seen.

The red dwarf star known as **Gliese 581** (about 20 light years from Earth in the constellation Libra) has a number of planets orbiting it. In 2007, it was announced that one of these planets, Gliese 581c, has a massive atmosphere that keeps the sunlight dim, though the temperature is roasting hot. This exoplanet could support liquid water, being in a position within the star's **habitable zone**. The planet is about 50 per cent larger than Earth, and has a mass five or six times that of Earth with about twice the surface gravity. Another planet in this same system, Gliese 581d, appears to be near the outer edge of the zone that could support life.

One newly discovered exoplanet, around the star **HD185269** in Cygnus, has some strange characteristics. The planet has a mass about equal to Jupiter, but it is only a fifth as far from its host star as Mercury is from the Sun. Being this close to its star means the planet should have a circular orbit, but astronomers have found the orbit to be highly eccentric (more like Pluto's orbit). The planet takes a fast 6.838 days to orbit its star.

The star **HD17156** is sun-like in size and located 250 light years away in the constellation Cassiopeia. In 2008, it was reported that the planet

Figure 14.7   The Lagoon nebula in the constellation Sagittarius is a region of ionised hydrogen clouds (red) some 5000 light years away. It is a large star-forming region. Maybe new planetary systems exist there as well.
(Photo J. Wilkinson)

Table 14.2 Some known exoplanetary systems

Star	Distance from Earth (light years)	Planets
Gliese 876	15	3
Gliese 581	20	3
HD69830	41	3
47 Ursa major	43	2
Upsilon Andromedea	44	3
55 Cancri	44	5
Mu Arae	50	4
HD190360	52	2
HD128311	54	2
HD82943	90	2
HD37124	108	3
HD12661	121	2
HD108874	223	2
HD73526	323	2

HD17156b orbits this star so close to its parent star (8 to 37 million kilometres), and absorbs so much heat that its night side glows permanently hot. The exoplanet takes 21.2 days to orbit the star (much longer than the previous record), and has a highly elongated orbit (eccentricity = 0.67).

Our solar system is not the only one with an asteroid belt. Astrophysicists using the Spitzer Space Telescope found evidence of a similar but much more massive belt around a star called **HD69830**, 41 light years away in the constellation Puppis. Collisions are thought to occur in this belt every 1000 years or so.

Note that according to the IAU definition of a planet, a planet must orbit a star, be massive enough to have a spherical shape, and have cleared the neighbourhood of its orbit of any other objects. For now, this definition only applies to our own solar system and to exoplanets. In recent years there have been reports of free-floating planetary-mass objects (ones not orbiting any star), sometimes called 'rogue planets' or 'interstellar planets'. For now, such objects are outside the working definition of 'planet'.

THE FUTURE

The decades-old search for other worlds like our Earth has been rejuvenated by the intense excitement and interest surrounding the discovery of planets surrounding other stars. However, there are many unanswered questions about the properties of exoplanets, such as details about their composition and the likelihood of their possessing moons. Another question is whether they might support life. Most of the exoplanets discovered so far are too close to their host star and too hot to be habitable: a habitable world is a solid-surfaced planet or moon that can maintain liquid water on its surface. The challenge now is to find terrestrial planets like Earth that are more likely to support life.

WEB NOTES

The Californian Institute of Technology has a good collection of NASA images from planetary exploration. Check out:
<http://pds.jpl.nasa.gov/planets/>
For further information on exoplanets and space missions check out:
<http://www.exoplanets.org>
<http://planetquest.jpl.nasa.gov/>

GLOSSARY

A ring	One of three prominent rings encircling the planet Saturn.
Albedo	A measure of the proportion of light reflected from a planet, asteroid or satellite.
Aphelion	The point in the elliptical orbit of a planet, comet or asteroid that is furthest from the Sun.
Apogee	The point in the orbit of the Moon or artificial satellite at which it is furthest from Earth.
Apparent magnitude	The visible brightness of a star or planet as seen from Earth.
Asteroid	A small rocky and/or metallic object, often irregular in shape, orbiting the Sun in the asteroid belt.
Asteroid belt	A large group of small bodies orbiting the Sun in a band between the orbits of Mars and Jupiter.
Asteroid number	The serial number assigned to an asteroid when it is discovered.
Astronomical unit (AU)	The mean distance between Earth and the Sun, about 150 million kilometres.
Atmosphere	A layer of gases surrounding a planet or moon, held in place by gravity.
Aurora	Curtains or arcs of light in the sky, usually in polar regions, caused by particles from the Sun interacting with Earth's or other planet's magnetic field.
Axial tilt	The angle between a planet's axis of rotation and the vertical; equal to the angle between a planet's equator and its orbital plane.
Axis	The imaginary line through the centre of a planet or star around which it rotates.
B ring	One of three prominent rings encircling the planet Saturn.
Big Bang	The event that caused the explosive birth of the universe about 13.7 billion years ago.
Black hole	An object with gravity so strong that no light or other matter can escape it.
C ring	One of three prominent rings encircling the planet Saturn.
Cassini division	Gap between Saturn's A and B rings.
Celestial equator	An imaginary line encircling the sky midway between the celestial poles.
Celestial poles	The imaginary points on the sky where Earth's rotation axis points if it is extended indefinitely.
Celestial sphere	An imaginary sphere surrounding Earth, upon which the stars, galaxies, and other objects all appear to lie.

Chromosphere   The layer of the Sun's atmosphere lying just above the photosphere (visible surface) and below the corona.

Cluster   A large group of stars or galaxies held closely together by gravitational attraction between them.

Coma   The diffuse, gaseous head of a comet.

Comet   A small body composed of ice and dust that orbits the Sun on an often highly elliptical path.

Constellation   One of 88 officially recognised patterns or groups of stars in the sky as seen from Earth.

Convection   A heat-driven process that causes hotter, less dense, material in a fluid to rise while cooler, denser material sinks.

Core   The innermost region or centre of a planet or star.

Corona   The tenuous outermost layer of the Sun's atmosphere, visible from Earth only during a solar eclipse.

Cosmology   The branch of astronomy that deals with the origins, structure and space–time dynamics of the universe.

Crater   A circular depression on a planet or moon caused by the impact of a meteor.

Crust   The surface layer of a terrestrial planet.

Dwarf planet   A celestial body that is in orbit around the Sun at a distance less than that of Neptune, has sufficient mass for its self-gravity to overcome rigid body forces so that it assumes a hydrostatic equilibrium shape (that is, it is nearly spherical), has not cleared its neighbourhood around its orbit, and is not a satellite.

Dwarf star   Any star of average to low size, mass and luminosity.

Earth   Our home planet, one of the major planets of the solar system.

Earthquake   A sudden vibratory motion of Earth's surface.

Eccentricity   A measure of how elliptical an orbit is. A perfect circle has an eccentricity of zero, and the more stretched an ellipse becomes the closer its eccentricity approaches a value of 1 (a straight line).

Eclipse   The total or partial disappearance of a celestial body in the shadow of another, such as solar eclipse, lunar eclipse.

Ecliptic   The apparent path of the Sun around the celestial sphere; also the plane of the orbit of Earth around the Sun.

Electromagnetic radiation   The type of radiation energy that includes radio, infrared, visible light, ultraviolet, X-rays and gamma rays.

Ellipse   The oval, closed loop path followed by a celestial body moving around another body under the influence of gravity.

Encke division   One of the narrow bands dividing Saturn's ring system. It is less prominent than the Cassini division.

Equator   The imaginary line around the middle of a celestial body, half way between its two poles.

Escape velocity   The minimum speed an object (such as a rocket) must attain in order to travel from the surface of a planet, moon or other body, and into space.

Exoplanet	A planet beyond the solar system, or belonging to another star. Most exoplanets found to date are massive giant planets like Jupiter. Also known as an extrasolar planet.
Galaxy	A huge group of stars, gas and dust held together by gravity and moving through space together.
Galilean moons	The four largest moons of Jupiter, discovered by Galileo Galilei in 1610. They are, in order of distance from Jupiter: Io, Europa, Ganymede and Callisto.
Gamma rays	Form of electromagnetic radiation with short wavelength and high frequency.
Gas giant	A large planet whose composition is dominated by hydrogen and helium. The gas giant planets in our solar system are: Jupiter, Saturn, Uranus and Neptune.
Gravity	The force of attraction between two masses.
Habitable zone	The region around a star at which a solid planet would be expected to maintain liquid water on its surface.
Heliosphere	A bubble blown into the interstellar medium by the pressure of the Sun's solar wind.
Ice	Solid state of water, and by extension methane or ammonia.
Inferior planet	A planet that is closer to the Sun than Earth.
Infrared radiation	Form of electromagnetic radiation with longer wavelength and lower frequency than the visible light region of the spectrum.
Ion	An electrically charged atom or molecule, either positive or negative.
Ionisation	The loss or gain by an atom or molecule of one or more electrons, resulting in the atom or molecule having a positive or negative electrical charge.
Ionosphere	A layer of Earth's atmosphere between 60 km and 1000 km above the surface, where a percentage of the gases are ionised by solar radiation.
Kepler's laws	Three laws discovered by Johannes Kepler that are used to describe the motion of objects in the solar system.
Kuiper belt	A region of the solar system beyond Neptune (between 30 AU and 50 AU from the Sun), contains icy and rocky bodies similar to those of the asteroid belt.
Lava	Molten rock flowing on the surface of a planet.
Light year	The distance light travels in one year, about 9.5 trillion kilometres. Used to give the distance to stars.
Luminosity	The amount of electromagnetic energy radiated by an object, such as a star. Luminosity depends on the temperature and surface area of the object.
Lunar phase	The appearance of the illuminated area of the Moon as seen from Earth.
Magnetic field	A region of force surrounding a magnetic object.
Magnetosphere	A region of space surrounding a planet or star that is dominated by the magnetic field of that body.

Magnitude	A measure of the brightness of a star or planet. The lower the magnitude the brighter the object.
Major axis	The longest diameter of an ellipse.
Mantle	The inner region of a planet that lies between its crust and its core.
Mare	A plain of solidified lava on the surface of the Moon; appears darker than the surrounding area.
Mass	The amount of matter in a body; usually measured in grams or kilograms.
Meteor	The bright streak of light that is seen when a rock or piece of space debris burns up as it enters Earth's atmosphere at high speed. Meteors that hit Earth's surface are called meteorites.
Meteoroids	Any small debris travelling through space; usually from a comet or asteroid.
Milky Way	The galaxy of stars and gas clouds that our solar system belongs to, seen as a luminous band of stars across the night sky. It is a spiral galaxy.
Moon	The only natural satellite of the Earth. The Moon takes about 28 days to orbit Earth once.
Natural satellite	A smaller heavenly body held in orbit around a bigger body by gravitational attraction. For example, the Moon is a natural satellite of Earth.
Nebula	A cloud of gas or dust, which may be illuminated by nearby stars.
Neutron	A subatomic particle with no electric charge.
Nuclear fusion	A process whereby light atomic nuclei (such as hydrogen or helium) combine to produce heavier nuclei, with the release of energy; often called 'burning'. Occurs in stars but not planets.
Occultation	The apparent disappearance of one celestial body behind another.
Oort cloud	A sphere of icy bodies surrounding the outer solar system. Much further from the Sun than the Kuiper belt.
Orbit	The path followed by one celestial body moving around another.
Perigee	The point in the orbit of the Moon or artificial satellite at which it is closest to Earth.
Perihelion	The closest point to the Sun in the elliptical orbit of a comet, asteroid or planet.
Period (of a planet)	The time taken for a planet to orbit the Sun.
Photosphere	The visible surface of the Sun or other star.
Planet	A celestial body that is in orbit around the Sun, has sufficient mass for its self-gravity to overcome rigid body forces so that it assumes a hydrostatic equilibrium shape (becomes nearly round), and has cleared its neighbourhood around its orbit.
Plutoid	A celestial body in orbit around the Sun at a distance greater than that of Neptune that has sufficient mass for its self-gravity to overcome rigid body forces so that it assume a hydrostatic equilibrium (near-spherical) shape, and that has not cleared the neighborhood around its orbit.
Prominences	Flame-like jets of gas thrown outwards from the Sun's chromosphere.

Protostar	A star in its early stages of formation.
Protosun	The part of the solar nebula that eventually developed into the Sun.
Pulsar	A rapidly rotating remnant from a supernova that emits bursts of energy at a regular rate.
Radio telescope	A telescope, often in the form of a dish-shaped receiver, designed to detect radio waves.
Radio waves	Electromagnetic waves of low frequency and long wavelength.
Red giant	A large, cool star of high luminosity that is in the later stages of its life.
Retrograde motion	The apparent westward motion of a planet with respect to background stars.
Retrograde orbit	The orbit of a satellite around a planet that is in the direction opposite to the rotation of the planet.
Revolution	The orbit of one body about another. One complete orbit is one revolution.
Revolution period	The time taken for one body to orbit another.
Rotation	The spin of a planet, satellite or star on its axis.
Satellite	Any small object (artificial or natural) orbiting a larger one, such as a moon orbiting a planet.
Scattered disc	A distant region of our solar system, thinly populated by icy minor heavenly bodies.
Seasons	The four divisions of the year of a planet whose axis of rotation is not perpendicular to the plane of its orbit; the distribution of solar radiation over the surface of the planet varies over its year. On Earth, the four seasons are summer, autumn, winter and spring.
Semi-major axis	The semi-major axis of an ellipse is equal to half the length of the long axis of an ellipse. The semi-major axis of a planetary orbit is also the average distance from the planet to its star.
Shepherd satellite	A satellite that constrains the extent of a planetary ring through gravitational interactions with the particles in the ring.
Sidereal time	The orbital period of a planet or satellite as measured with respect to the stars.
Solar flare	A sudden release of energy in or near the Sun's corona, resulting in a burst of radiation and particles being emitted into space.
Solar nebula	The cloud of gas and dust from which the Sun and solar system formed.
Solar system	The Sun, planets and their satellites, asteroids, comets and other related objects that orbit the Sun.
Solar wind	A stream of charged particles or ions emitted by the Sun.
Space probe	A spacecraft or artificial satellite used to explore other bodies (such as the planets or Moon) in the solar system. Such a craft contains instruments to record and send back data to scientists on Earth.
Space station	A craft or vehicle that is in a stable orbit around Earth or other planet and is the temporary home of astronauts.
Spectrometer	A device used to analyse the light from stars.
Spectroscopy	The analysis of light from a planet or star to determine the composition and condition of the planet or star.

Star	A self-luminous sphere of gas.
Sunspot	A highly magnetic storm on the Sun's surface that is cooler than the surrounding area and so appears dark compared to the rest of the Sun.
Superior planet	A planet that is more distant from the Sun than Earth.
Supernova	An exploding star, which briefly emits large amounts of light.
Synchronous orbit	A condition in which a moon's rotation rate and revolution rate are equal.
Tectonic forces	Forces within a planet or moon that lead to the deformation of the crust of the body.
Terrestrial planet	A planet whose composition is mainly rock (Mercury, Venus, Earth and Mars).
Transit	The passage of one astronomical body in front of another, for example when a planet passes in front of the Sun's disc as seen from Earth.
Weight	The force of gravity acting on an object.
White dwarf	A small, dense, very hot but faint star. The final state of all but the most massive stars before they fade out.
X-rays	Electromagnetic radiation with short wavelength and high frequency (between ultraviolet and gamma rays).
Zodiac	Name given to a group of twelve constellations that lie along the path followed by the Sun across the sky.

INDEX